How to Build LGBTQ+ Inclusive Workplaces

Bringing together the latest research with practical insights from the authors' professional experience, this important book provides a context for the conversations that are needed within organisations and offers practical guidance towards action that can be taken to improve the working life of LGBTQ+ employees.

The book begins by asking how we got here. It outlines the development of stigma towards the LGBTQ+ community from both a historical and psychological perspective before going on to explore the ways in which societal attitudes manifest in the work environment. It then looks specifically at LGBTQ+ experiences in the workplace, covering discrimination and exclusion and their impact at both an individual and organisational level before taking an intersectional view of LGBTQ+ identity, and particularly how it interacts with race, disability and age. The book then provides clear and practical guidance on how to build an LGBTQ+ inclusive workplace, covering organisational policy and culture, leadership and allyship. Throughout, the authors use case studies to demonstrate how to implement policies across a range of regions and offer strategies to minimise homophobic and discriminatory attitudes.

Taking a psychological approach to this important topic, the book is essential reading for all those looking to build and sustain welcoming and inclusive workplaces across all sectors. It will also be of interest to students in psychology, management and human resources studying workplace attitudes and culture.

Binna Kandola is a co-founder and senior partner at Pearn Kandola, a practice of business psychologists. He is a visiting professor at Leeds University Business School, UK, and has spent over 40 years researching diversity, inclusion and bias in the workplace. As a practitioner, he has worked with some of the world's leading organisations, including American Express, Microsoft, NATO and the World Bank. He has been on the Asian Power List for the last five years.

Ashley Williams is a business psychologist who specialises in LGBTQ+ diversity and inclusion within organisations. Ashley's PhD research explored the impact of stereotypes on the career experiences of LGBTQ+ individuals. During this time, she attended international research summer schools and presented her research at conferences in the UK and across Europe. At Pearn Kandola, Ashley works closely with organisations to support them in developing strategy regarding LGBTQ+ inclusion and designs and delivers training initiatives to help raise awareness of this topic and the role of allies.

How to Build LGBTQ+ Inclusive Workplaces

A Psychological Approach

Binna Kandola and Ashley Williams

Routledge
Taylor & Francis Group

LONDON AND NEW YORK

Designed cover image: zhengshun tang via Getty Images

First published 2025
by Routledge
4 Park Square, Milton Park, Abingdon, Oxon OX14 4RN

and by Routledge
605 Third Avenue, New York, NY 10158

Routledge is an imprint of the Taylor & Francis Group, an informa
business

British Library Cataloguing-in-Publication Data
A catalogue record for this book is available from the British
Library

ISBN: 978-1-032-78867-8 (hbk)
ISBN: 978-1-032-78862-3 (pbk)
ISBN: 978-1-003-48958-0 (ebk)

DOI: 10.4324/9781003489580

Typeset in Sabon
by Apex CoVantage, LLC

To a best friend and father: Jim Williams

Contents

Acknowledgements

From the moment we discussed the idea of this book to its publication, the UK has seen four prime ministers, two heads of state and the first pandemic in over 100 years. On a personal level, we've witnessed two weddings, the birth of a daughter and the arrival of a grandson.

A lot can happen in three years! The process of writing this book is a reflection of the times we've lived through. Most of the conversations, discussions and debates that shaped these pages took place through electronic communications. In many ways, we existed in our own bubble, relying on each other to bring this project to completion.

We must also recognise the courage of the authors who have provided histories of human sexuality, allowing us to understand better the true diversity of human experience. Writing about this topic has often been a dangerous and challenging undertaking, but their work has been invaluable. Their research has led to a more enlightened understanding of human sexuality, paving the way for a more inclusive and progressive future.

We are immensely grateful for the support of our partners, Martin and Jo, who stood by us throughout. Our colleagues at Pearn Kandola deserve special thanks for their continued encouragement.

A heartfelt thank you to Paul May, whose insights and feedback were invaluable in shaping the book. We are also deeply appreciative of our editors at Routledge, Zoe Thomson-Kemp and Hannah Rich, for their unwavering support, encouragement and, above all, patience throughout this endeavour.

This book is a testament to the collaborative spirit and resilience of everyone involved. Thank you.

Foreword

In 2024, the landscape of LGBTQ+ inclusion in the workplace is one of celebration, visibility and, in many ways, triumph. Across the UK and the Republic of Ireland, over 180 Pride events are set to take place, involving communities, companies and individuals in a colourful and joyous expression of identity, solidarity and resilience. For organisations, Pride is more than just a moment on the calendar; it has become a statement of values. Participation in public events and hosting internal celebrations for employees are not only encouraged but also expected in many sectors. Rainbow lanyards, wristbands and badges, which symbolise support for LGBTQ+ rights, have become part of the fabric of workplace culture. They signify an organisation's commitment to inclusion, offering employees and the public an overt sign that they stand on the side of equality.

This display of allyship is no longer a bold or controversial move; in fact, it's often a badge of corporate pride. Many organisations now prominently display their inclusion awards and accolades, whether they are ranked in the lists of inclusive employers or have achieved certification from other diversity and inclusion benchmarks. Such displays represent a sea change in attitudes towards LGBTQ+ individuals. The punishments for discriminating, bullying or excluding someone based on their sexual orientation are now more severe than ever, reinforced by workplace policies and legal protections. Across many countries, including the UK, employment rights for LGBTQ+ individuals are enshrined in law, and same-sex marriage continues to be legalised in more nations worldwide.

This progress is indeed worth celebrating, as it marks an unprecedented and significant shift in societal attitudes towards sexual minorities. But while these advances are clear, they are also relatively new. Just a quarter of a century ago, the landscape was very different. In the late 1990s and early 2000s, organisations faced significant risks in openly supporting LGBTQ+ inclusion, even if they were inclined to. In the 1980s, UK local authorities who dared to include sexual orientation as part of their drive on equality would be labelled "Loony Left" and as a danger to society. This was a time when homophobia, both subtle and overt, was not only more widely accepted but also often institutionally enforced. People who identified as LGBTQ+ were stigmatised, marginalised and discriminated against, both overtly and through everyday practices of exclusion.

At that time, even the idea of allyship—a term we use frequently today to describe someone who supports LGBTQ+ individuals despite not being part of the community—was far from mainstream. To publicly align oneself with LGBTQ+ causes was often seen as radical or rebellious. Stereotypical portrayals of LGBTQ+ individuals in the media further compounded this stigma. Gay men were often depicted as effeminate, comedic figures; lesbians were portrayed as either overly masculine or hypersexualised; and bisexual individuals were depicted as confused or untrustworthy. Beyond these representations, LGBTQ+ characters were frequently portrayed as villains. It was not uncommon for LGBTQ+ characters to be cast as antagonists in films and television, reinforcing a societal narrative that equated LGBTQ+ identities with deviance or immorality.

This historical context is important because it illustrates how fast things have changed. In less than three decades, society has moved from an era where LGBTQ+ inclusion was largely ignored, if not outright resisted, to a time where it is actively embraced by many organisations.

Yet, it is crucial to acknowledge that while progress has been made, this transformation is incomplete. The public display of support for LGBTQ+ rights—whether in the form of rainbow flags, corporate sponsorship of Pride events or the adoption of inclusion policies—is a step forward. But it does not mean that the workplace, or society at large, is free of the challenges that LGBTQ+ individuals face.

The Legacy of Homophobia in Organisational Structures

As we reflect on the progress made, we must recognise that the obstacles facing LGBTQ+ individuals in the workplace are deeply rooted in history. Much of the homophobia that persists today in certain regions of the world can be traced back to colonialism. During the age of European imperialism, colonial powers imposed legal codes that criminalised homosexuality on their colonies. These laws were a direct import from the moral and religious views of the colonisers, who often saw same-sex relationships as sinful or unnatural. This colonial legacy continues to shape attitudes towards LGBTQ+ individuals in many countries today, particularly in Africa, the Caribbean and parts of Asia.

It is sobering to realise that the same European countries that once spread homophobic laws across the globe are now among the most progressive in advocating for LGBTQ+ rights. The irony is that these nations, which once persecuted their own LGBTQ+ populations and exported those prejudices to other parts of the world, now criticise those same regions for their lack of progress on LGBTQ+ inclusion.

This complicated dynamic reminds us that the global struggle for LGBTQ+ rights is deeply intertwined with histories of power, oppression and cultural imperialism.

In organisational settings, the remnants of these historical prejudices still influence workplace culture and structures. While many companies have made significant strides towards fostering more inclusive environments, others lag behind, either due to cultural resistance, organisational inertia, or fear of backlash. In some cases, organisations may adopt a superficial commitment to LGBTQ+ inclusion, promoting diversity and equality externally but failing to address the internal biases and barriers that LGBTQ+ employees face. This phenomenon, often referred to as "rainbow-washing" or "pink-washing," highlights the gap between performative support and genuine inclusion.

Psychological Impacts of LGBTQ+ Exclusion

As organisational psychologists, we are particularly interested in the impact that exclusion—or even perceived exclusion—can have on individuals in the workplace. Decades of research in social and organisational psychology have shown that feeling excluded or marginalised can have profound effects on an individual's mental health, job performance and overall well-being, and it is no different for LGBTQ+ employees in workplaces.

Conversely, workplaces that promote LGBTQ+ inclusion tend to see higher levels of employee engagement, job satisfaction and productivity among their LGBTQ+ workforce. The psychological safety that comes from knowing one is accepted and valued allows individuals to perform at their best, to collaborate more openly with colleagues and to contribute their unique perspectives to the organisation. This is beneficial not only for the individual but also for the organisation as a whole. Diverse teams have been shown to be more innovative, creative and adaptable, and fostering a culture of inclusion can provide a competitive advantage in today's global marketplace.[1]

But the journey towards creating truly inclusive workplaces is far from over. As much as we would like to believe that we live in the most accepting and progressive era in history, the reality is more complex. While legal protections for LGBTQ+ individuals have expanded in many countries, cultural attitudes often lag behind. Homophobic and transphobic behaviour, both overt and subtle, still occurs in workplaces around the world. The celebration and enjoyment of the progress that has been achieved can make us complacent to the challenges that remain.

These challenges are exacerbated for LGBTQ+ employees who hold other marginalised identities, such as people of colour, or individuals with disabilities, and older people. Intersectionality—the idea that people can experience multiple, overlapping forms of discrimination—complicates the experiences of LGBTQ+ individuals in the workplace. For instance, a Black lesbian woman may face not only homophobia but also racism and sexism, which can interact in ways that make her workplace experience uniquely difficult.

The Role of Organisations in Advancing LGBTQ+ Inclusion

Given these challenges, what can organisations do to create more inclusive workplaces for LGBTQ+ individuals? First, it is important to recognise that inclusion is not just about policy—it is about culture. While having non-discrimination policies and employee resource groups (ERGs) are important, they are not sufficient on their own. Organisations must also work to change the underlying attitudes, behaviours and structures that can perpetuate exclusion.

One way to foster a more inclusive culture is through education and training. Many organisations have implemented unconscious bias training to help employees recognise and address their own biases. While this is a good first step, it is also important to go beyond awareness and focus on behaviour change. For example, training programmes can teach employees how to be effective allies, how to use inclusive language and how to challenge discriminatory behaviour when they see it. Leadership plays a critical role in setting the tone for the rest of the organisation. When senior leaders demonstrate a commitment to inclusion—whether by sponsoring LGBTQ+ ERGs, speaking out against discrimination or mentoring LGBTQ+ employees—it sends a powerful message to the entire organisation that diversity and inclusion are valued.

Another key strategy is to promote visibility. Many LGBTQ+ individuals in the workplace still feel the need to hide their identity, either because they fear discrimination or because they believe it will harm their career prospects. When LGBTQ+ employees see others like them in leadership positions, it can help alleviate these fears and create a sense of belonging. Organisations can support LGBTQ+ visibility by celebrating LGBTQ+ role models, both within the organisation and in the wider community.

Finally, organisations must ensure that their inclusion efforts are intersectional. This means recognising and addressing the unique challenges faced by LGBTQ+ employees who belong to multiple marginalised groups. By adopting an intersectional approach, organisations can create more nuanced and effective inclusion strategies that meet the needs of all employees.

Looking Ahead: The Future of LGBTQ+ Inclusion

As we look to the future, it is clear that while progress has been made, there is still much work to be done. The legal landscape for LGBTQ+ rights continues to evolve, with both new victories and emerging challenges. For instance, while same-sex marriage is now recognised in many countries, there are still regions where LGBTQ+ individuals face harsh discrimination, and where their rights are actively being rolled back. At the same time, even in countries with progressive legal frameworks, there are ongoing struggles related to the workplace, from subtle signals to outright exclusion.

This book has been written with a clear intention: to provide both individuals and organisations with practical tools and insights to navigate these complexities. Our aim is not merely to celebrate the progress that has been made but to examine critically where challenges remain and how they can be addressed. Readers will gain an understanding of the historical context that has shaped LGBTQ+ rights, the psychological impacts of inclusion and exclusion and the organisational strategies that can foster genuine inclusivity. We offer evidence-based insights drawn from organisational psychology to help readers recognise the difference between performative support and meaningful change, and how to cultivate an environment where all employees can thrive, regardless of their sexual orientation or gender identity.

Ultimately, this book is designed to inspire action. We hope it will spark important discussions, encourage introspection and provide clarity on how we can all contribute to building more inclusive and equitable workplaces. Whether you are an individual seeking to better understand the experiences of LGBTQ+ colleagues or a leader committed to transforming your organisation, this book will offer valuable guidance, provoke thoughtful debate and empower you to be an advocate for inclusion. By shining a light on the ongoing struggles and triumphs of the LGBTQ+ community in the workplace, we hope to help create spaces where everyone, regardless of identity, can truly belong.

Note

1 Donovan, P. (2020). *Profit and Prejudice: The Luddites of the Fourth Industrial Revolution* (1st ed.). Routledge. https://doi.org/10.4324/9781003098898

Part 1

How We Got Here

Chapter 1

Introduction

For many people, 14th February 2014 was Big Bang Day as far as gender and gender identity were concerned. On that Valentine's Day users of the world's biggest social media network, Facebook, would be able to select from a list of 58 terms to identify their gender rather than being restricted to male or female. The social media giant said in a statement on its diversity page

> When you come to Facebook to connect with the people, causes, and organisations you care about, we want you to feel comfortable being your true, authentic self. An important part of this is the expression of gender, especially when it extends beyond the definitions of just "male" or "female." So today, we're proud to offer a new custom gender option to help you better express your own identity on Facebook.[1]

The company had "collaborated with our Network of Support, a group of leading LGBT advocacy organizations, to offer an extensive list of gender identities that many people use to describe themselves."

While other social media organisations had made similar changes before Facebook, this was the one that was covered most in the traditional media and which captured people's attention. The comments beneath the statement reflect a range of views: from those welcoming it to those challenging its validity and meaning with the tone varying from civility to abusive. These debates, and the ways in which they are conducted, continue today. A significant shift had occurred, however, by bringing the classifications into public consciousness.

In this chapter we will introduce you to some, but by no means all, of the terms often used to define LGBTQ+ identities and also provide a brief description of the structure of the book.

Who Are the LGBTQ+ Community?

The LGBTQ+ community encompasses individuals from multiple identity groups, relating to a person's sexuality, sex or gender.

DOI: 10.4324/9781003489580-2

Here we will shed some light on what we mean by these terms, as well as who the acronym is specifically referring to. You may sometimes see longer terms, for example, LGBTQIA, but no matter the length, it will almost always be followed by a plus sign (+) which is a recognition that there are several identities which are not covered by the acronym.

Regardless of its length, it can be argued that no term can ever encompass the entire spectrum of gender and sexual expression. Language evolves and, in this area, it can seem to do so at great speed. Not only does the language change but also its meaning and its tone. Sometimes the meanings of words can become more negative over time, a process known as pejoration, or can become more positive, that is, amelioration. Using today's powerful computing resources it's possible to look at the movement in the meanings and the words which typically accompanied their use over the decades.[2] Lesbian, for example, became more widespread in the 1960s, and its initial neighbours were both positive ("sweet-faced", "nice-looking" "demure") and negative ("nymphomaniac" "unfaithful" "bitchy").[3] By the turn of the millennium the words most closely connected with lesbian were more neutral, including "legalising rights," "activist" and "stigmatised."

Three phases of movement and development of the word gay were identified in the analysis. The first phase is in the 1860s, when it was connected to words denoting a positive emotional state "merry, blithe, cheerful, and mirthful."[4] Over a hundred years later it starts to be associated with homosexuality: "bawdy," "flirtatious" and "gullible." By the 1980s it has become more negative: "incest," "promiscuous," "high risk" and "homosexual."

Queer has also been on its own process of change similar to that of gay. In the 1860s the words most associated with it included "droll," "funny," "strange" and "odd." By the 1970s it was still linked with funny or strange but also with "weird," "ponce" and "faggy." It became even more negative in the 1990s: "nasty," "miscreant" and "lecher." By the 2000s it had as its nearest neighbours more positive words such as "motherly," "reverent" and "mocking."

Homosexual has moved its nearest neighbours from the 1930s which included "recidivist," "schizophrenia" and "phobias" to the 1970s "perverts," "promiscuous," "deviates" and "rape" to the 2000s "legalising," "bigoted" and "guiltless." Today there are still negative words as neighbours, including "immorality" and "promiscuity," but also more neutral ones, including "same-sex" "activist" and "rights." In the 1940s the nearest neighbours to be found with bisexual were primarily negative and medical "chromosome", "manic-depressive", and "psychopath". By the 2000s this has changed to "stereotyped" "misperceptions" as well as "effeminate" and "prostituted".

The analysis gives us an indication of attitudes. While the words that are most associated with sexual orientation are less judgemental and critical than in the past, it's not nearly as positive as we might like to think—a central theme in this book and why we decided to write it.

It's all but certain the words people use to describe gender identity and sexuality will continue to evolve. This means we are going to come across terms and identities that we're not familiar with, which is why it is important to continue to educate ourselves on this evolving topic, rather than deciding not to engage in a discussion at all. It's an understandable reaction because we won't want to embarrass ourselves by saying something that might offend.

However, it's not helpful to self-censor, and by doing so, failing to provide support to others when they need it or acting as an ally when it's required. Apart from this list, we have opted to define the words in the context in which we use them in each chapter rather than providing a full glossary. We provide descriptions of some terms below, but this is not intended to be comprehensive, and it's always important to bear in mind that even if the words don't change, their meanings may alter over time.

Defining LGBTQ+

A person's sexuality or sexual orientation refers to the sex, gender or genders to which they are typically sexually or romantically attracted.

The majority of the population identify as heterosexual, which refers to individuals who are sexually or romantically attracted exclusively to people of the other sex (i.e., men who are attracted to women, and women who are attracted to men). The LGBTQ+ acronym refers to many individuals who do not identify as heterosexual, including those who identify as Lesbians, which refers to women who are sexually or romantically attracted only to women, and Gay people, meaning either men who are sexually or romantically attracted only to other men, or women who are sexually or romantically attracted only to other women.

B stands for Bisexual, meaning a person with sexual and romantic attraction to more than one gender.

The T in the acronym stands for Transgender. This is an umbrella term for people whose gender identity and/or gender expression differ from what is typically associated with the sex they were assigned at birth. For those whose gender identity is associated with the sex they were assigned at birth, the term "cisgender" is increasingly used.

This community includes individuals who identify as transgender men and transgender women, as well as those who identify as non-binary. A person who is non-binary doesn't identify strictly as a man or a woman, and they may use alternative identity markers to identify themselves.

The Q in the acronym stands for Queer. Queer is an umbrella term, which is used to describe sexual and gender identities that are not heterosexual and not cisgender.

Sometimes, for those who identify as queer, the terms "lesbian," "gay" and bisexual are perceived to be too limiting or fraught with cultural connotations they feel do not apply to them. Queer was once considered a derogatory term but has been reclaimed by some LGBTQ+ people to describe themselves. However, it is not a universally accepted term even within the LGBTQ+ community, so caution is advised when using it outside of describing the way someone self-identifies.

The Debate Around Sex and Gender

There are different views about these terms and what they really mean and over time this discussion has become extremely polarised. While many people won't necessarily have strong thoughts on this matter, we are now in a position where voicing any perspectives, concerns or questions in relation to this topic has left individuals open to attack from those who identify with a different school of thought.

Any view that is shared, or concern that is raised, is met with hostility and "cancelling" by those with another view. This is stifling conversation and is limiting organisations' progress in creating safe and inclusive environments for all employees.

Dr. Hilary Cass, in carrying out a review of gender identity services in the UK, found the same difficulties.[5] She noted a number of things in conducting the review. "The toxicity of the debate is exceptional," and one consequence of this is that professionals are "afraid to openly discuss their views, where people are vilified on social media, and where name-calling echoes the worst bullying people" (p13). The other point of relevance here regards the evidence base. "This is an area of remarkably weak evidence, and yet results of studies are exaggerated or misrepresented by people on all sides of the debate to support their viewpoint" (p13). The same applies to dealing with these issues in the workplace.

Therefore, the first thing we need to do is focus on how we can have constructive, meaningful and respectful conversations, where different considerations are encouraged and heard, and dealt with to prevent negative implications against any groups of employees now or in the future. If we can't safely have those conversations, we're at risk of this culture becoming more toxic and less progressive. To move forward, it is imperative that we create safe spaces where we can discuss this openly, educate ourselves on this topic and hear different viewpoints.

In the United Kingdom, gender reassignment is a protected characteristic under the Equality Act 2010. Decisions in tribunals have established that this also includes protection for non-binary and gender-fluid people. Gender-critical beliefs are also protected under the same legislation, which means that organisations have to balance the rights of people on all sides of the argument. Organisations cannot take action which are seen to be favouring one side over the other. According to legal experts at Shoosmiths, employers need to be careful not to end up "favouring the protection of one group of people with a protected characteristic over another, as opposed to carefully balancing competing rights and assessing whether parties have acted reasonably in the circumstances."[6]

By listening to different perspectives, organisations will have to find a way of incorporating different needs. The bottom line here is that it is complicated, or it certainly still feels that way given that we just don't have enough research at the moment regarding the implications of certain practices that aim to advance transgender inclusion. A lack of research and evidence is leaving businesses feeling stuck—what really is best practice? How can we ensure the safety and respect of everybody at work?

As practitioners of business psychology, we deal with the challenges that businesses are currently faced with. Whatever your view on gender identity, it is a current and very real challenge that society faces in terms of what it means to be male or female and one that needs the engagement of not just policy makers and activists but researchers from a variety of disciplines. In the meantime, organisations have individuals who identify as transgender and non-binary, but who are not feeling safe, included or respected at work. It is also true that we are in a society where women experience discrimination, exclusion, violence and abuse. It is equally important that we ensure that safety for one group of people does not compromise it for another. In taking action, however, we also need to consider broader implications before jumping into what feels like the "right" decision or what is advised as "best practice" based on the information we have today.

Our position as business psychologists, which will be evident throughout this book, is that there is insufficient research and data to reach definitive conclusions or decisions right now. But what we will do is provide guidance for how to manage through difficult decisions and approach discussions with openness and respect for one another.

Structure of the Book

The book is in three parts:

Part I: How We Got Here
Part II: LGBTQ+ in the workplace
Part III: Building an LGBTQ+ inclusive workplace

Part I: How We Got Here

Chapter 2: The Long, Dark Journey from Acceptance to Exclusion

It's very easy to assume that the attitudes that we hold today are continuation of those that have always been held. This chapter examines whether the assumptions we make are true.

We explore the attitudes that were held in ancient history in different parts of the world.

We look at how and why attitudes towards those who are not hetero-sexual started to change, in particular examining the role of religion.

Chapter 3: Love, Law and the Couch

This chapter highlights how psychology and psychiatry came to replace that of the church and the role these disciplines played in the process of seeing homosexuality as a deviation. The damage that was caused is still being felt today. We describe the key scientific studies which challenged these partial and stigmatising views of LGBTQ+, including the key work of Alfred Kinsey, the anthropologists Ford and Beach, and psychologist Evelyn Hooker.

Part II: LGBTQ+ in the Workplace

Chapter 4: Stereotypes and Stigma at Work

This chapter explores the ways in which the societal attitudes explored in Chapters 2 and 3 manifest in the work environment. We discuss stereotyping of the LGBTQ+ community, the stigmatisation that results and the impact this has on individuals in terms of their own self-image. Stereotypes associated with the different identities are described so that we can examine the ways in which these groups differ in the ways that they are viewed.

We also look at some of the things that can be done to counter stereotyping and stigmatisation.

Chapter 5: Discrimination and Exclusion at Work

This chapter provides an overview of the research exploring LGBTQ+ experiences at work, highlighting the extent of their discrimination and exclusion. We look at the discrimination that occurs at different points in the employee life cycle from recruitment and selection through to promotion. We also look at how societal attitudes and the desire to avoid discrimination and work in inclusive workplaces impact career choices at a young age.

Chapter 6: The Impact of Exclusion

This chapter explores the impact of the previously outlined experiences of exclusion on LGBTQ+ individuals. This includes a discussion on the con-cealment of identity, and the negative impact this has on an individual's self-image, wellbeing and performance.

Chapter 7: The Venn Diagrams of LGBTQ+ intersectionality: Disability, Race and Age

This chapter takes an intersectional view of LGBTQ+ identity and looks at the unique experiences that emerge from overlapping identities. Three areas will be described in detail: LGBTQ+ and race, LGBTQ+ and disability, LGBTQ+ and age or more specifically older people.

Part III: Building an LGBTQ+ Inclusive Workplace

Chapter 8: The Evolving Organisation

Here we look at how legislation and attitudes differ around the world and the difficulties that this can create for global organisations. Organisational silence on LGBTQ+ is a particular obstacle that we discuss in more detail.

Chapter 9: Organisational Action to Be LGBTQ+ Inclusive

This chapter focuses on actions that organisations can take to be more LGBTQ+ inclusive.

We provide the overall framework and structure for taking action which can be built on in the future and create cultures within organisations to enable everyone to feel more authentic in the workplace to the benefit of individuals, organisations and societies.

Notes

1 www.facebook.com/photo.php?fbid=567587973337709&id=105225179573993 &set=a.196865713743272

2 Shi, Y., & Lei, L. (2020). The Evolution of LGBT Labelling Words: Tracking 150 Years of the Interaction of Semantics with Social and Cultural Changes. *English Today*, 36(4), pp. 33–39.

3 Shi, Y., & Lei, L. (2020). The Evolution of LGBT Labelling Words: Tracking 150 Years of the Interaction of Semantics with Social and Cultural Changes. *English Today*, 36(4), pp. 33–39.

4 Shi, Y., & Lei, L. (2020). The Evolution of LGBT Labelling Words: Tracking 150 Years of the Interaction of Semantics with Social and Cultural Changes. *English Today*, 36(4), pp. 33–39.

5 https://cass.independent-review.uk/wp-content/uploads/2024/04/CassReview_ Final.pdf

6 www.shoosmiths.com/insights/articles/protecting-gender-reassignment #:~:text=Following%20a%20case%20in%20the,whose%20identity%20 fluctuates%20at%20different

Chapter 2

The Long, Dark Journey from Acceptance to Exclusion

The banquet was becoming more boisterous as the evening wore on. The wine flowed and the guests—the most noted intellectuals of Greek society—competed with one another to come up with the most compelling theories about the nature of love. Late in the evening, it was finally the turn of the comedian and playwright Aristophanes to make his contribution. Tipsy and hiccupping, Aristophanes deliberated on the origins and evolution of human beings, who he declared were originally the most beautiful objects that the gods had created, being cast in the shape of a circle, the most perfect shape in existence.

With two heads, four arms, four legs and two sets of sex organs, the two halves of a person could be male and male, female and female, or male and female. But a growing sense of superiority, fuelled by hubris and ambition, led the humans to make a serious mistake: they decided to challenge the gods. The reaction of the deities was swift, furious and decisive. They slashed humans in half, leaving each with just one head, one set of sex organs, two arms and two legs.

Since that divisive moment, Aristophanes continued, each individual has been looking for its perfect other half in order to make itself whole again. The person's missing half could be of the opposite or the same sex and, once found, would make them complete.[1]

This legend, both brutal and charming, makes no judgement about the sex of one's partner—what matters is finding the person who makes you complete. When a previously "double man" finds his natural other half, "they are wonderfully overwhelmed with affection and intimacy and love, and never wish to be apart for a moment."[2]

This chapter explores the shifting perceptions of same-sex attraction and bisexuality, tracing their transition from natural facets of human experience to being labelled as sinful and illegal and ultimately defining aspects of identity. We begin by recounting origin stories, myths and legends, focusing on how diverse sexualities were depicted, including the sexuality of renowned generals. From there, we analyse how political power has historically used sexuality as a means of control and coercion. With the increasing enforcement

DOI: 10.4324/9781003489580-3

of religious doctrines, legal measures were introduced to criminalise same-sex relationships. Finally, we examine how LGBTQ+ history has been systematically erased to reinforce the idea that such behaviours are abnormal and deviant.

Origin Stories and Myths

Some people may look at the Greeks' attitude to sex as "weird." David Halperin, a professor of the history and theory of sexuality, having researched this area extensively, concluded:

> what I, and many others, have learned from this work is that it is not the Greeks who were weird about sex, but rather it is we today, particularly men and women of the professional classes, who have a culturally and historically unique organisation of sexual and social life and, therefore, have difficulty understanding the sex/gender systems of other cultures.[3]

It is hard for us to understand the attitudes towards sex of other cultures and, indeed, of people of the past because we view them through our own cultural lenses, making judgements based on our present standards. Imagine, for example, how a member of the audience felt at the first performance of Hamlet. We put ourselves in their shoes and might think they felt special, filled with the recognition that they were witnessing a momentous event. However, our knowledge of the lasting significance of the play gets in the way of understanding how those attending the premiere felt because the one thing we can say for certain is that they didn't feel this sense of historical occasion at all.[4]

The same limitations affect the way we view sex and sexuality in earlier times, when today's labels and categories did not exist. The idea that some things were normal and others deviant wasn't even a way of thinking, and origin stories give us an insight into how early societies viewed sexuality.

Theories of how humans originated and evolved appear in many ancient cultures including North America, Japan, China and the Middle East.[5] Ancient Indian texts include stories of births which have no connection with heterosexuality, and a number of religions have deities who can move between being male and female. Other origin stories include cross-dressing, gender transformation, third genders, men presenting as women and men giving birth. The South Indian goddess Yellamma, still worshipped today, possesses her followers' bodies and can change their sex.[6]

One story form of Greek myth involves an older person—a warrior, teacher or a master of some kind—having a young boy as an apprentice or a pupil, with the tale of Zeus and Ganymede among the better known. Zeus spots the shepherd Ganymede tending his flock on a mountainside and struck by his beauty, is determined to take him as his lover. Adopting the form of an eagle, Zeus swoops down and abducts the shepherd to Mount Olympus, where he gives him the esteemed position of cupbearer. In doing so he enrages Hera, the goddess of women and marriage, whose daughter Hebe had been replaced as cupbearer. Eventually Ganymede gives up the role to the anger of Zeus himself. Zeus recognises that he has treated the boy badly and immortalises Ganymede by representing his beauty in the constellation of Aquarius.[7]

Other surprising mythical stories providing powerful intimations of same-sex relationships include King Arthur and the Knights of the Round Table and Beowulf, which were systematically ignored by the translators of these Anglo-Saxon works. It is only with more recent translations that these interpretations have been explored.[8]

Evidence of Attitudes to Same-Sex Love in Ancient Times

An important source of evidence about same-sex sexuality among men and women in the ancient world is art and literature.

In ancient Greece, sexuality and procreation were not connected in the way they are today. If you were to have children, then it was expected that this would be done within a marriage, but sexual pleasure could be obtained outside of the marriage with men as well as women.[9] Having looked at many examples of ancient Greek poetry, Spencer reached the conclusion that "in all this it is clear that homosexual love was considered to be far superior to heterosexual love."[10] Because it didn't have the motivation of procreation, it could be seen as pure love. As the artwork of the time shows, same-sex relationships were part of everyday life and there are many examples of Greek men openly sharing their admiration and love for other men.[11] This includes Plato and his love for Aster, as well as Alcibiades, who appears in Plato's *Symposium* and who attempted unsuccessfully to seduce Socrates.

While in Greek society it was expected that men would engage in same-sex relationships, there were differences in the attitudes and culture across the three regions of Sparta, Athens and Lesbos.

Seventh-century BCE Sparta was a militaristic state which placed great emphasis on the skills and strengths of its warriors. While people lived communally, they were separated by sex up to the age of 30 and it was not unusual for men to have sex with other men. Marriage to women was permitted at the age of 18, and the custom in Sparta was that the bride would present as a likeness of a man, cutting their hair and wearing male attire.[12]

Spartan women had a reputation, certainly among the Athenians, of enjoying having sex with other women. Elite women would have slaves and so were not expected to carry out many household chores and duties. Instead, they were to keep themselves fit and healthy in order to carry out their most important role, bearing children. Sexual relationships between women were considered acceptable because they didn't interfere with their primary task.

In the Athenian world, women's lives were much more controlled and restricted. Life was far less communal for them and they would be expected to live in their husband's house, confined to the area allocated to women. Being separated from the men, and living in close proximity to other women, it's easy to see that relationships and sexual ties would develop between them.[13]

Our sources for attitudes about sexuality in ancient art and literature are mainly the products of men. Levels of literacy among women and girls were low, and so their thoughts and experiences were not generally recorded and preserved unlike those of men. One very significant female figure who wrote about her experiences was the poet Sappho. Born on the island of Lesbos about 612 BCE, her role seems to have been akin to that of a headteacher of a community of girls in a premarital school. Through her poetry, we learn that women engaged in sexual relationships with other women, and as Leila Rupp says, Sappho's was a "lonely voice in the record of women desiring women in ancient worlds."[14] Her influence can be seen in the way that she is seen as the reference point for same-sex relationships between women over the centuries. In England in the 18th century, a pamphlet blamed Sappho for bringing into the world "A new Sort of Sin." An 11th-century Muslim poet in Spain earned herself the title of the "Arab Sappho."[15] For many generations, the name given to same-sex love between women was Sapphism.

Ancient Chinese myths and artwork also give us an insight into attitudes towards same-sex love. In one story, the king's lover Long Yang becomes upset when he imagines the king discarding him for another more attractive man, in the same way that he would discard a smaller fish for a larger one. The king makes it clear that he would never do this and issues a decree that no one should be allowed to refer to a more beautiful man in his presence. If they did so, the penalty would be execution of the person's family. Long Yang would later be the name associated with same-sex male love.[16]

Roman astrological texts give us additional clues about how same-sex love was regarded. The theory was that the alignment of the stars and planets at the time of birth determined someone's sexuality. In other words, Fate was the determining factor. Poetry, for example by Catullus, provide explicit demonstrations of love-making between men.[17] Despite this, same-sex desire between women was thought to be unnatural and considered to be transgressive behaviour on a level with adultery and prostitution.[18]

Freud wasn't the first person to try and interpret our dreams. In the 2nd century CE ancient Egyptians were trying to understand sexual dreams. These analyses clearly reflect the attitudes of their time, so while sex between men is seen as acceptable and even natural, the opposite is true for sex between women. Dreams revealing sex between women were seen as omens foretelling some bad or sad outcome. For women, even imagining same-sex love was to be avoided.

The evidence shows that throughout the ancient world same-sex relationships took place and without the stigma that is attached to them today. In what is now Peru, pre-Inca Mochica or Moche pottery depicts a wide range of sexual behaviour, including masturbation and homosexuality.[19] Evidence of same-sex relationships between women can be found in the pottery and texts of both the Mochica (100–800s CE) and the Chimu (1100–1400s CE). People who didn't conform to gender norms were noted, but they were described in an objective way, expressing curiosity more than disapproval:

> She is a woman who has a foreskin, she has a penis. She is a possessor of arrows; an owner of darts . . . she has a manly body . . . she often speaks in the fashion of a man; she often plays the role of a man. She possesses facial hair.

Choice of companions is also remarked upon: "She is a possessor of companions, one who pairs off with women . . . . She has sexual relations with women, she makes friends with women. She never wishes to be married."[20]

Ancient Hindu artworks show people engaged in same-sex activity, which was clearly considered acceptable and worthy of displaying in temples. The same is true for Japanese Buddhism, where sex between men was sometimes seen as necessary as that between men and women. As with ancient soldiers in Europe, the samurai also engaged in same-sex relationships.[21] Japanese culture was influenced by the beliefs and norms of Chinese society, including their attitudes and practices regarding same-sex love. Apart from bestiality and incest, legal codes had very little to say about sexual practices. One of the most revered of the religious figures is Kūkai, who came to be known after his death as Kōbō Daishi or "The Great Teacher." Legend has it that this celebrated figure introduced male love to Japanese culture after he had returned from a period in China. The acceptance of homosexuality was so widespread that the Jesuit missionary Saint Francis Xavier was shocked to see that people engaged in this sin, as he saw it, so openly.[22]

It is generally thought that the Aztecs had laws which punished homosexual acts, and as such, they were different from other native peoples. The Florentine Codex was written in Nahuatl in the 16th century by Bernadino de Sahagun with a team of indigenous people and shows that the Aztecs may have been done a disservice in this regard, as they were more accepting than was previously thought and definitely not as cruel as the Spanish conquistadors in their attitudes towards same-sex relationships.[23] Other evidence shows that ancient societies in Peru, China and India, along with Greece, were aware of women engaged in sexual relationships with other women.

Military Leaders and Their Armies

A Google search on books about one of the greatest military commanders ever, Alexander the Great, produces approximately 3,600,000 results. His leadership abilities have been a constant source of fascination for hundreds of years, and he has been an inspiration to many generals since his death to the present day. And he was almost certainly bisexual. He wouldn't have been called this back in his day, not because people were afraid to but simply because that identity had not been created. He was living his life the way that he wanted and was neither indulged nor criticised because of it.[24]

As opposed to the ancient Greeks, Roman politicians took the opportunity to criticise opponents for their sexual conduct. Julius Caesar was one of a number of Roman emperors who was believed to have had relationships with men as well as women. His sexual energy was such that it was said of him, by the consul Curio the Elder, that he was *omnium virorum mulier, omnium mulierum virum* or "every woman's man and every man's woman."[25] Caesar was ridiculed for his lack of sexual restraint and for being the more submissive partner when with other men, but not for his choice of partner. He did not conform to the fashion of how men should present themselves and by choosing to take great care over his appearance, by dressing unconventionally for a man (wearing his tunic both loosely belted and with sleeves) and by depilating himself he knew that this would be considered to be effeminate.[26]

These criticisms never really left their mark on the intended victims and Caesar's assassination had nothing to do with his sexual activities. Attempts were also made to tarnish the reputations of many people but with no success, and as Louis Crompton surmises in his book *Homosexuality and Civilisation,*

> If allegations of homosexuality had ended careers in Rome—as they would have, certainly, in eighteenth- or nineteenth century England—the Roman political stage during the turbulent last century of the republic would have been bereft of Sulla, Pompey, Cataline, Caesar, Clodius, Mark Anthony, and Octavius; in short, it would have been deprived of most of its principal players.[27]

Emperor Hadrian's love for Antinous was so great that on his early death, the emperor memorialised him by creating a city in his name, Antinoöpolis. A popular religious cult which revered Antinous as a god lasted several centuries before being supplanted by Christianity.

As with the leaders, so with their followers. Soldiers in conquering armies would rape the defeated combatants to show their absolute strength and superiority over them. In this case, men having sex with other men was clearly a demonstration of domination, humiliation and subjugation of weaker people. But this doesn't explain why soldiers in the same army would be encouraged to make love to one another. One rationale is the lack of women in the battalions led to an excess of sexual energy which had to be expended on other men. However, it has been argued that same-sex relationships helped to increase soldiers' attachment, affection and loyalty to one another so that they would then fight harder to protect their colleagues. According to psychologist Julia Shaw, "the Spartans in ancient Greece encouraged homosexuality among elite troops"[28] for exactly these reasons. The evolutionary biologist Robin Dunbar supports this line of reasoning: "they had the not unreasonable belief that individuals would stick by and make all efforts to rescue other individuals if they had a lover relationship."[29]

Diplomatic protocols supported the same line of reasoning. King Richard I of England, otherwise known as Richard the Lionheart, not only spent long periods away from his wife on the crusades to the Middle East but we also know that he and King Philip II of France shared a bed and were probably in a relationship. In a 2008 article in *The Guardian*, mediaeval historian Helen Castor explains that "Richard's decision to share a mattress with Philip was the ultimate public demonstration of trust" and that these practices are still reflected in the language "when we talk of sealing a deal." In the same article another historian suggests that this was merely "an accepted political act, nothing sexual about it." In other words, this situation was nothing to do with Richard being anything other than heterosexual.[30]

Arguing that many men will only have sex with other men when there are no women around, or that it is a show of dominance or a requirement of diplomatic etiquette, can make it seem as if we are trying to deny the possibility of homosexuality. But this inference may simply reveal present-day attitudes where people must be one thing or the other, heterosexual or homosexual. Julia Shaw argues that if a heterosexual willingly engages in sex with someone of the same sex, even if only occasionally, this surely means they are bisexual.[31]

Religion, Power and Politics: Weaponising Sexuality

By the turn of the first millennium these accepting attitudes towards sexuality were already changing. Religions were moving from having female gods to male ones and religious authorities were increasingly taking a much more judgemental and restricted view of sex, including their approach to same-sex sexuality.

A key moment presaging this change occurred in 309 CE when the Church Council of Elvira determined that sexual behaviour was now of interest and concern to the church. This was followed up a few years later with the proclamation by Emperor Constantine that Christianity would now be the religion of the Roman Empire. Sexual behaviour now is of interest to the state as well as the church. Something that had always been treated as private became regulated and legislated against.[32]

Penitential manuals began to appear from the 6th-century listing sins and the acts the sinner or penitent was to carry out to redeem themselves. The 7th-century manual of Theodore and Bede's of the 8th century included sins committed by women having sex with one another. Theodore's *Penitential* (*Paenitentiale Theodori*) specified a penance of three years if the convicted women were not married or widowed. Masturbation carried the same level of punishment, with married women being given higher penalties. Bede, in his *Penitential (Paenitentiale Bedae)*, increased the period of penance for nuns using an instrument to seven years. (It isn't entirely clear whether the penalty was increased because the women were nuns or because they were using aids.) Hincmar of Reims, writing in 9th-century France, was particularly concerned about the use of tools and devices. It's clear that relationships between women were known about across the centuries, even if we can only learn about them indirectly through the writings of censorious men.[33]

As its power increased, the Catholic church sought to eliminate those who could challenge its authority including people of other faiths, Protestants, heretics, witches and homosexuals. The changes in attitudes led by the church solidified gradually from around 500 CE to 1500 CE. The church's strategy for maintaining its dominance centred on drawing clear differences between its own teachings and the beliefs of those competing sects, to undermine them. These other sects, they argued, adopted sinful and deviant behaviour such as allowing women to play a prominent role and acceptance of homosexuality.

For example, the Catholic church accused the Cathars or Albigensians—an ascetic sect originating in Bulgaria which gained followers throughout Europe—of having orgies and permitting same-sex relationships. It is believed that the term "bugger" originated by association with the movement's Bulgarian origins.[34]

Religious beliefs also provided a useful smokescreen for political agitation against powerful figures and groups. The Knights Templar were originally created to protect pilgrims on their way to the Holy Land. Devoted to their task, they adopted the monastic way of life and kept themselves separate from the rest of society, eventually becoming a wealthy and powerful order, and by 1300 it owned 870 castles and houses, with properties stretching from Ireland in the west to Palestine in the east. Their power was seen as a challenge to the authorities and their wealth envied. Add to this mix a financially strapped monarch in Philip IV of France, who was facing resistance to his tax-raising efforts at home, and the Knights Templar represented a way of solving a political predicament. He and a key minister, Guillaume de Nogaret, accused the Knights Templar of all manner of unholy acts including apostasy, "carnal relations" between men as well as worshipping cats.[35] Some 2,000 people were arrested on this pretext in the early morning of 13 October 1307,[36] and with further pressure from the Pope, the 200-year order was dissolved in just five years. This was a key moment as it demonstrated that even the most powerful could not escape the attention of a vengeful monarch who could weaponise religious doctrine.

The medieval theologian Thomas Aquinas had a particularly important part to play in bridging the gap between religion and legislation. He had four separate categories for sinful sex: masturbation, bestiality, sex in an unnatural position and sex with someone of the same sex—and here he identifies not only men having sex with other men but women having sex with other women.[37] As this could only be for pleasure it was both sinful[38] and a contravention of natural law, a view that found its way into legal frameworks.[39]

The tactics used by the Catholic church to discredit other sects were then used by the Protestant reformers against it. Luther, Calvin and Zwingli were highly critical of the Catholic church's teachings, leadership and lack of morality. Places of entertainment were shut down, including alehouses, theatres and arenas for cockfighting and bearbaiting.

When it came to sexual relationships, the Puritans shared the same views as Thomas Aquinas but were more determined to ensure that these doctrines were followed. This led to an important shift from judging the behaviour of individuals to judging the people themselves.

Crime and Punishments

Imagine going about your business and coming across a man standing on the balcony of his house swearing and blaspheming. Obviously drunk, he then proceeds to unzip himself, dip his penis into his beer and then splashes the drink onto the crowd that has now assembled. What would he be guilty of in your opinion, if anything? Eccentricity? Being abusive? Antisocial behaviour? Whatever label you have applied we are pretty certain you won't have said sodomy. In the first Elizabethan period, this was what he was found guilty of.[40]

Around 1300 CE the word "sodomy" came into use—and with it a new crime. Sodomy however was a category of behaviour and could encompass a wide range of activities including Christians having sex with Jews, Muslims or people of other faiths. An increasing number of states introduced legislation making sodomy a crime. The Spanish Inquisition made such acts punishable by being burnt at the stake.[41] In England, 13th-century laws made sex with Jews, children and members of one's own sex punishable by being buried alive.

Records from Perugia in Italy show that penalties there varied by age: those between 12 and 15 years of age could expect three months in jail, those over 15 would face jail and a fine, and adults who didn't pay their (higher) fines could be tied, paraded naked through the streets, beaten and then exiled.[42]

In 1270, a document from France recommended punishments for men and women committing sodomy. First-time male offenders would be castrated, a second offence would lead to them losing their penis, and a third offence would result in burning. A similar set of punishments was described for women although they make rather confusing reading: for the first two offences women would lose their "member," and for the third offence they would be burned.[43]

However, these discriminatory doctrines weren't applied uniformly across Europe. The Spanish Inquisition might have been burning people at the stake, but other states took a very different view. For example, after the German Empire conquered Sicily in 1231 the new legal code covered many typical topics such as heresy and usury but made no reference to sexual behaviour. Neither did all legal codes made sex between women unlawful, England being a case in point. However, in France, Spain, Italy, Germany and Switzerland women could be sentenced to death for having sex with other women.[44]

Even where discriminatory laws were in place, they weren't necessarily fully implemented. Serbians, Russians and Bulgarians adopted Christianity later than most other European countries, between the 9th and 11th centuries. As was the norm at the time, sex between men was seen to be as evil as adultery and bestiality, carrying a penance of 15 years, and kissing between men carried a penance of 40 days. These prohibitions, however, were apparently ignored. A member of the English Queen Elizabeth's court returning from Russia where they witnessed romantic relationships between men said that this was the result of "savage soil, where the laws do not bear sway."[45]

The city states of Italy are emblematic of a pattern that has developed from that time until today as they fluctuated between general acceptance and periods of hostility. In Renaissance Florence increasing the fines was found to have such little impact on the behaviour of its people that they were slashed by 80% for a first offence from 50 florins to 10.[46] There were only 50 prosecutions for sodomy in Florence between 1348 and 1461. Of these cases, ten people received the death sentence. A similar situation prevailed in 14th-century Venice and Genoa.[47]

Women's sexuality was wrapped up in the craze to identify witches in the early modern period. Accusations of witchcraft was one way that men had of controlling and abusing women and sexual deviance was taken as a sign that someone was a witch. Artwork from 16th-century Germany for example show witches as independent from men, usually naked, and engaging in highly erotic and sexualised behaviour.[48]

There was intellectual resistance to the imposition of one form of sexuality to the exclusion of all others. It became apparent to Renaissance students of ancient Greek and Roman cultures that they adopted very different attitudes to sex. Marsilio Ficino (1422–99), who translated Plato, stated: "the reproductive drive of the soul being without cognition, makes no distinction between the sexes [and] is naturally aroused for copulation whenever we judge anybody to be beautiful."[49]

On a trip to Rome in 1580 the French essayist Montaigne saw men being married to one another at the Church of Saint John.[50] Yet at the same time artists of the period were being charged with sodomy, including Leonardo da Vinci, Sandro Botticelli, Benvenuto Cellini and Giovanni Antonio Bazzi.[51] It is typical of the struggles that continue today with resistance being shown in different ways to the bigotry that people were facing.

There has also been much speculation as to whether Shakespeare was gay. Scholars are always keen to point out that we shouldn't make assumptions about Shakespeare's life based on his body of work. Nevertheless, the sonnets are seen as love poems and most of them are in fact addressed to a man. The most famous of the sonnets, commonly read at marriage or civil partnership ceremonies and starting with "Shall I compare thee to a summer's day?" was written to a man. Leading Shakespearean scholar Professor Sir Stanley Wells said that "it's not unreasonable to suppose that Shakespeare was sometimes in a relationship, as we say, with men and also with women other than his wife."[52]

During the 17th century the term "masculine love" came into the language to describe men who loved men, thus making a distinction between different types of sexualities.[53] Analysis of the language of the period reveals that the phrase "male conversation" was a more direct reference to sex between men.[54] These expressions can be contrasted with the more judgmental and ambiguous use of the word "sodomy."

A concerning development was the globalisation of these rules by their application to the overseas colonies being established, which began the process of overturning long-held attitudes about the diversity of sexual conduct in those societies. In 1607, in Jamestown, the first English North American settlement sodomy, rape and adultery were all declared as sins punishable by death. In 1655 in the colony of New Haven, lesbianism was added as a crime.[55] The attitudes of this period have clearly had an impact which is still being felt today.

Nevertheless, with the restoration of the monarchy in England the attitudes of the pre-Republican era also returned and it was again seen as acceptable for somebody to be engaged in relationships with someone of the same sex. When William and Mary became king and queen after the 1689 abdication of James II, prevailing attitudes to relationships were summed up by the king who said that "it seems to me a most extraordinary thing that one may not feel regard and affection for a young man without it being criminal."[56] Queen Anne's relationships with women were well known and were transferred to the movie screen with Olivia Colman's Oscar-winning portrayal in *The Favourite* (2018).

The increasingly hostile and antagonistic attitudes inevitably had an impact on those being ostracised. From 1709 onwards there came the development of what became known as Molly houses (possibly linked to the Latin *mollis*, used to describe someone who didn't fit the masculine appearance of a man).[57] These were secret, private clubs where men were able to meet and express themselves more fully. At least half of the people who went to the Molly houses (referred to as Molly men) cross-dressed. The Molly men attracted the disapproval of the authorities who seemed determined to punish them for their behaviour. The penalties for those who were convicted of sodomy ranged from being sent to the pillory to fines, imprisonment (from six months to three years) and the death sentence. It has been estimated that an average of one Molly per week was put in the stocks in the two decades between 1720 and 1740. This barbaric treatment of a community of people continued for a century.

Increasingly the Mollies were being referred to as hermaphrodite, effeminate men. The same became true of women who loved women, and the courts were able to assign them the label of hermaphrodite which they had to accept if they wished to remain out of trouble.

These pressures led to people being less open about their sexuality and establishing greater numbers of secret communities. This was especially the case in the growing cities but people had to be very careful because being involved in this underground world exposed them to potential blackmail and ruin.

Erasing LGBTQ+ from History

The 2004 movie *Alexander*, a biopic of Alexander the Great, is not considered to be one of Oliver Stone's best films, failing to generate much of a return on its estimated $155m budget. Nevertheless, it did get to number one in the Greek box office, partly due to an unexpected publicity boost when a group of 25 lawyers tried to get the movie banned because they objected to the portrayal of Alexander as gay. The leader of the protesters explained that "this is not an attack against gays, but rather a demand to make the film historically accurate."[58]

But the movie was, as we noted earlier. Historically correct: Alexander is known to have had sexual relationships with men and was almost certainly gay if not bisexual. The fact that some people think the reputations of great historical figures have been tarnished when details like these are shared reveals the stigma which is still attached to those who are not heterosexual. It seems we prefer our revered figures to be both exceptional and yet conventional. However, these judgemental attitudes were non-existent when Alexander lived, and this is a recent and blatant attempt by the lawyers to maintain the fiction that he was heterosexual.

In order to rid the world of anyone involved in same-sex relationships, the change in attitudes had to be accompanied by attempts to alter, adapt and remove any historical records which suggested that the heroic figures of the past were anything other than heterosexual. Knowing that ancient societies accepted the diversity of peoples' sexual orientation would have totally undermined their inhuman project.

Many of the celebrated people that we have referred to in this chapter have had aspects of their biography omitted to ensure that their heterosexuality could never be questioned. John Addington Symonds, the 19th-century poet, literary critic and historian, researched the lives of the ancient Greeks. Today we would refer to him as a closeted gay man, and as such he wanted to share the lives of the ancient Greeks to change the prevailing highly intolerant attitudes of the time. As we will see in the next chapter, he did make his essays available to other researchers, but his collected papers, including those relating to his personal life, were eventually sealed in the British Library. His daughter was given access to the papers in 1949, but she kept secret the nature of their contents. This represents a double erasure one of his life and being, and the other of the historical facts he had uncovered neither of which fitted the tide of the times.[59]

The one consolation we can take from this is that at least the papers were only hidden from view and not burnt. This can't be said for some early archaeologists and collectors of Mochica pottery who considered these sexual representations to be obscene and a form of pornography and destroyed them.[60]

Literary works were subtly altered to ensure there were no references to same-sex love and relationships. Legends of Beowulf and of King Arthur were edited to ensure that there were no signs of effeminacy in their descriptions of what the Victorians considered to be prototypical masculine figures. As Caitlin Rimmer suggests there was a "deeper investment" in the presentation of their heterosexuality. Given the colonisation that was taking place at the time, it was important that all-conquering figures like the Anglo-Saxon Arthur were presented unambiguously so that sexual deviations could be seen to belong to those cultures and races which were inferior.

Penitential texts were not translated from Latin so that people could not see the type of acts people were engaged in that needed to be punished. When they were translated there were significant omissions with sections relating to female same-sex acts and the use of devices and tools.[61]

Michelangelo's poetry was published by his nephew in 1623, 59 years after the artist's death. It took over a quarter of a millennium, in 1892, to discover that poems apparently about women were in fact about men: the pronouns had been changed from masculine to feminine, phrases changed or cut, and notes added[62] to ensure that future generations would not see Michaelangelo as gay.[63]

Similarly, any discussion about Shakespeare could not include questions about his sexuality which meant failing to bring attention to the fact that many of the sonnets were written to a man. *As You Like*, a play with much gender confusion, ends with a wedding. Rosalind, who has spent much of the play in disguise as the male Ganymede, is told, "'thou mightst joyne his hand with his,/Whose heart within his bosome is'." As the Shakespearean expert Emma Smith points out,

> Joining two male hands together doesn't seem quite the image of heterosexual union that editors want to see. All "straighten" one of the "his" into "her". Thus the edited text is at once more normative, more heterosexual, and more prescribed than the first printed version in the Folio.[64]

The poems of Catullus also had their explicit lines removed and it's remarkable to realise that English schoolboys in Shakespeare's time, including the young William himself, "would have learned far more about homosexuality from his classroom reading than a student in the age of Kinsey."[65]

The significant individuals and texts mentioned earlier in this chapter are just a small part of a broader, systematic process of heterosexualisation. This is not the work of a few, but rather a collective effort carried out by countless individuals over an extended period, each contributing, often in seemingly minor ways. These seemingly small actions, however, have had a profound cumulative effect. They have collectively erased a community's perspective on history, denying LGBTQ+ individuals the understanding that their feelings are not only normal but that there were times in the past when such feelings would have been accepted without question.

The public shaming of those who were not conforming to society's ideals of "normal" sexual conduct hastened the process of people choosing to hide an essential part of their identity from others.

In San Francisco, 1970, just four years after the Compton's Cafeteria riots and one year after the Stonewall Inn uprising, the streets bore graffiti that read: "If God had wanted homosexuals, he would have created Adam and Freddy." Though crude, this message was seized by anti-gay activists and, over time, sharpened into the slogan that still echoes today: "It's Adam and Eve, not Adam and Steve." An origin story twisted and weaponised, its goal was to marginalise and erase a vibrant, diverse community.

But where history was altered to oppress, it is now being restored to empower. Thanks to the dedication of historians and activists, we are recovering the voices and stories that were nearly silenced, ensuring a future where inclusivity and understanding can thrive. As we reclaim these narratives, we move closer to a world where every individual can exist freely and authentically.

Up until this point, people were judged primarily by what they did, but the focus soon shifted to who they were, as labels began to emerge. These labels, which would go on to define and differentiate people based on their sexual orientation, will be discussed in the next chapter.

Key Points

The way human actions, particularly around sexuality, are judged depends heavily on the time and context in which those actions occur. Throughout much of history, and in many parts of the world, human sexuality was recognised in various forms, a fact that was often accepted without much scrutiny or debate.

Today, we emphasise concepts like acceptance and inclusion—terms that imply a conscious, active effort to embrace different identities. However, this modern language may not accurately capture earlier attitudes, which were often natural, organic, and accepted without explicit thought or need for articulation.

Over centuries, political opportunism, religious extremism and oppressive laws reshaped societal views on sexuality. These forces brought us to where we are today: a world in which LGBTQ+ individuals are stigmatised and marginalised.

Such discriminatory attitudes and regulations were then exported to regions where sexuality had been far less policed. Meanwhile, history itself was rewritten, and cultural artefacts were destroyed, all to reinforce heterosexuality as the only "normal" form of relationship. Same-sex relationships were further weaponised by being linked to racist ideologies, marking them as behaviours associated with so-called "inferior" races.

Notes

1 Destrée, P., & Giannopoulou, Z. (eds.). (2017). *Plato's Symposium: A Critical Guide*. New York, NY: Cambridge University Press.
2 Mondimore, F.M. (1996). *A Natural History of Homosexuality*. Johns Hopkins University Press, p. 10.
3 Halperin David, M. (2002). *How to Do the History of Homosexuality*. Chicago University of Chicago Press, p. 3.
4 Adapted from Halperin David, M. (2002). *How to Do the History of Homosexuality*. Chicago University of Chicago Press.
5 Rupp, L.J. (2009). *Sapphistries: A Global History of Love between Women*. NYU Press.
6 Divine by Lucinda Ramberg June 3rd 2019 The Immanent Frame: Secularism, Religion and the Public Frame. https://tif.ssrc.org/2019/06/03/divine/. Accessed 26th August 2023.
7 Calimach, A. (2007). The Exquisite Corpse of Ganymede: A Cursory Overview of an Ancient Gender Studies Discourse. *Boyhood Studies*, 1(2), pp. 117–137.
8 Wade, E. (2022). Skeletons in the Closet: Scholarly Erasure of Queer and Trans Themes in Early Medieval English Texts. *Elh*, 89(2), pp. 281–316. https://doi.org/10.1353/elh.2022.0011
9 Mondimore, F.M. (1996). *A Natural History of Homosexuality*. Johns Hopkins University Press.
10 Spencer, C. (1996). *Homosexuality: A History*. United Kingdom: Fourth Estate, p. 43.
11 Aldrich, R. (2004). Homosexuality and the City: An Historical Overview. *Urban Studies*, 41(9), pp. 1719–1737. www.jstor.org/stable/43201476
12 Rupp, L.J. (2009). *Sapphistries: A Global History of Love between Women*. NYU Press.
13 Rupp, L.J. (2009). *Sapphistries: A Global History of Love between Women*. NYU Press.
14 Rupp, L.J. (2009). *Sapphistries: A Global History of Love between Women*. NYU Press, p. 33.
15 Rupp, L.J. (2009). *Sapphistries: A Global History of Love between Women*. NYU Press, p. 1.
16 Wu, J. (2003). From 'Long Yang' and 'Dui Shi' to Tongzhi: Homosexuality in China. *Journal of Gay & Lesbian Psychotherapy*, 7(1–2), pp. 117–143.
17 Taylor, R. (1997). Two Pathic Subcultures in Ancient Rome. *Journal of the History of Sexuality*, 7(3), pp. 319–371.
18 Rupp, L.J. (2009). *Sapphistries: A Global History of Love between Women*. NYU Press.

19 Wołoszyn, J.Z., & Piwowar, K. (2015). Sodomites, Siamese Twins, and Scholars: Same-Sex Relationships in Moche Art. *American Anthropologist*, 117(2), pp. 285–301. www.jstor.org/stable/24476214

20 Rupp, L.J. (2009). *Sapphistries: A Global History of Love between Women*. NYU Press, pp. 26–27.

21 Sato, H. (2018). Forbidden Colors: Homosexuality has Traditionally been Accepted in Japan, but the Country's Laws have yet to Catch Up. *World Policy Journal*, 35(1), pp. 49–53.

22 Crompton, L. (2003). *Homosexuality and Civilization*. Cambridge, MA: Belknap Press of Harvard University Press.

23 Rupp, L.J. (2009). *Sapphistries: A Global History of Love between Women*. NYU Press.

24 Jonsson, M. (2014). The Sexuality of Alexander the Great: From Arrian to Oliver Stone. *Wittenberg History Journal*, 43, p. 13.

25 Olson, K. (2014). Masculinity, Appearance, and Sexuality: Dandies in Roman Antiquity. *Journal of the History of Sexuality*, 23(2), pp. 182–205.

26 Olson, K. (2014). Masculinity, Appearance, and Sexuality: Dandies in Roman Antiquity. *Journal of the History of Sexuality*, 23(2), pp. 182–205.

27 Halperin, D.M. (2002). *How to Do the History of Homosexuality*. Chicago: University of Chicago Press.

28 Shaw, J. (2022). *Bi: The Hidden Culture, History, and Science of Bisexuality*. New York: Abrams Press, p. 83.

29 www.nationalgeographic.com/science/article/homosexual-animals-debate

30 www.theguardian.com/uk/2008/mar/19/monarchy.france. Accessed 2nd October 2023.

31 Shaw, J. (2022). *Bi: The Hidden Culture, History, and Science of Bisexuality*. New York: Abrams Press.

32 Mondimore, F.M. (1996). *A Natural History of Homosexuality*. Johns Hopkins University Press.

33 Rupp, L.J. (2009). *Sapphistries: A Global History of Love between Women*. NYU Press.

34 Rupp, L.J. (2009). *Sapphistries: A Global History of Love between Women*. NYU Press.

35 Crompton, L. (2003). *Homosexuality and Civilization*. Cambridge, MA: Belknap Press of Harvard University Press, p. 193.

36 Spencer, C. (1996). *Homosexuality: A History*. United Kingdom: Fourth Estate.

37 Rupp, L.J. (2009). *Sapphistries: A Global History of Love between Women*. NYU Press.

38 Milhaven, J.G. (1977). Thomas Aquinas on Sexual Pleasure. *The Journal of Religious Ethics*, 5(2), pp. 157–181. www.jstor.org/stable/40017725

39 Marcin, R.B. (1998). Natural Law, Homosexual Conduct, and the Public Policy Exception. *Creighton Law Review*, 32, p. 67.

40 www.shakespeare.org.uk/explore-shakespeare/podcasts/lets-talk-shakespeare/was-shakespeare-gay/

41 Hepple, J. (2012). Will Sexual Minorities Ever be Equal? The Repercussions of British Colonial "Sodomy" Laws. *The Equal Rights Review*, 8, pp. 50–64. (NHH p. 24).

42 Spencer, C. (1996). *Homosexuality: A History*. United Kingdom: Fourth Estate.

43 Rupp, L.J. (2009). *Sapphistries: A Global History of Love between Women*. NYU Press.

44 Rupp, L.J. (2009). *Sapphistries: A Global History of Love between Women*. NYU Press.

45 Spencer, C. (1996). *Homosexuality: A History*. United Kingdom: Fourth Estate, p. 133.

46 Crompton, L. (2003). *Homosexuality and Civilization*. Cambridge, MA: Belknap Press of Harvard University Press.
47 Spencer, C. (1996). *Homosexuality: A History*. United Kingdom: Fourth Estate.
48 Rupp, L.J. (2009). *Sapphistries: A Global History of Love between Women*. NYU Press.
49 Spencer, C. (1996). *Homosexuality: A History*. United Kingdom: Fourth Estate, p. 134.
50 Spencer, C. (1996). *Homosexuality: A History*. United Kingdom: Fourth Estate.
51 Ledain, D. (2019). *This Forbidden Fruit: Male Homosexuality, a Culture and History Guide*.
52 www.shakespeare.org.uk/explore-shakespeare/podcasts/lets-talk-shakespeare/was-shakespeare-gay/
53 Cady, J. (1992). 'Masculine Love,' Renaissance Writing, and the 'New Invention' of Homosexuality. *Journal of Homosexuality*, 23(1–2), pp. 9–40.
54 Forker, C.R. (1996). 'Masculine Love,' Renaissance Writing, and the 'New Invention' of Homosexuality: An Addendum. *Journal of Homosexuality*, 31(3), pp. 85–93. https://doi.org/10.1300/J082v31n03_06
55 Spencer, C. (1996). *Homosexuality: A History*. United Kingdom: Fourth Estate.
56 Spencer, C. (1996). *Homosexuality: A History*. United Kingdom: Fourth Estate, p. 180.
57 Olson, K. (2014). Masculinity, Appearance, and Sexuality: Dandies in Roman Antiquity. *Journal of the History of Sexuality*, 23(2), pp. 182–205. https://muse.jhu.edu/article/542474
58 https://edition.cnn.com/2004/LAW/11/25/alexander/
59 Mondimore, F.M. (1996). *A Natural History of Homosexuality*. Johns Hopkins University Press.
60 Wołoszyn, J.Z., & Piwowar, K. (2015). Sodomites, Siamese Twins, and Scholars: Same-Sex Relationships in Moche Art. *American Anthropologist* 117(2), pp. 285–301. www.jstor.org/stable/24476214
61 Rimmer, C. (2023). On Queer Remembering and Misremembering: Exploring Queer Aphasia. *Journal of Folklore Research*, 60(2), pp. 37–66. https://dx.doi.org/10.2979/jfr.2023.a912088
62 Ledain, D. (2019). *This Forbidden Fruit: Male Homosexuality, a Culture and History Guide*.
63 www.metmuseum.org/articles/james-saslow-interview-michelangelo-poetry
64 Smith, E. (2007). *The Cambridge Introduction to Shakespeare (Cambridge Introductions to Literature)*. Cambridge University Press, p. 60.
65 Crompton, L. (2003). *Homosexuality and Civilization*. Cambridge, MA: Belknap Press of Harvard University Press.

Chapter 3

Love, Law and the Couch

Homosexuality was invented in 1869 by the journalist, translator, author and human rights activist Karl Maria Kertbeny. A Hungarian émigré living in Germany, Kertbeny wrote an open letter to the Minister of Justice arguing that the proposed new penal code should not make same-sex love between consenting adults a crime. Ultimately motivated by the tragic death of a close friend, Kertbeny introduced the word "homoszexualitás" to the world—and ignited a mixture of fear and oppression which continues to the present day.

Kertbeny's friend committed suicide in 1840 after being blackmailed for his sexual orientation. Seeking to understand the forces at work in his friend's persecution, Kertbeny came into contact with a community which was so hidden he referred to it as a "sect."[1] He hoped that his new coinage—*homoszexualitás*—would not only help to identify gay people more clearly but also lead to understanding and empathy. He saw himself as an ally to this clandestine community and claimed he was "normally sexed" and of "normal sexuality." This insistence only raised question marks about his sexual orientation.[2] His argument that homosexuality in both men and women was natural and therefore could not and should not be considered deviant, sinful or criminal was rejected. The legislators he sought to influence used his novel term to make homosexuality categorically illegal.

Creating a Homophobic Society

To create a racist society requires collaboration, explicit or implicit, among philosophers, scientists and lawmakers. Together they can create a framework which dehumanises groups of people, create hierarchies of worthiness and institutionalise the denigration and ultimately criminalisation of sections of society based on an identity. Anthropologists Audrey and Brian Smedley described the way in which this can happen to create racist societies,[3] but the same principles and procedures can be applied to the mistreatment of many other groups.

The creation of a binary divide between heterosexuals and homosexuals led to the labelling, negative stereotyping and stigmatisation of the latter

DOI: 10.4324/9781003489580-4

group. What used to be disapproval of things that people did, changed into the demonisation of who people were. Scientists lent a hand by producing spurious evidence to support the derogation of the group, evidence which could then be used to support legislation to authorise discrimination against them.

As soon as definitional dividing lines are created, justificatory explanations follow. Love between women was explained by the claim that these women formed a distinct category of human: the hermaphrodite. This notion was used in court to account for and excuse their behaviour. Men who adopted the more submissive role in a relationship did so because their minds had been corrupted.[4] The binary regime created strict, logical laws for sexual orientation: "One was thereby conceptualised as both a man and either a heterosexual or homosexual, a woman and either a heterosexual or a lesbian."[5]

Other social changes reinforced the binary framework. These included separation of the gender roles with the establishment of the female homemaker and the male breadwinner. White middle-class women were not expected to go out to work—that was the role of working-class women and women of colour. Men and women were together because they had complementary roles, and this was the natural state of affairs. Anything that disturbed this notion was to be resisted and challenged.

Tellingly, women's social lives also became separate from those of men, creating opportunities for women to form friendships with one another without social disapproval. There is evidence of very passionate relationships developing during this period. Long-term, monogamous relationships between women who were not financially dependent on men blossomed in New England and came to be known as Boston marriages.[6] The Ladies of Llangollen ran away from their aristocratic homes in 1778 to live together in north Wales and are celebrated as role models in the LGBTQ+ community.[7]

Making a Science of Sex

To make a science, you need names for the objects you intend to study. As we have seen Karl Maria Kertbeny came to win the war of words with "homosexuality" although it wasn't his chosen battle. The British sexologist Havelock Ellis called the term a "barbarous neologism sprung from a monstrous mingling of Greek and Latin stock"[8]—though you could say the same about "television." Surprisingly, it wasn't until 1976 that homosexuality entered the Oxford English Dictionary.

Whatever your attitudes to the acronym LGBTQ+, it's worth considering that we could be grappling with UUDUDV or even MWUDUDV+. Karl Heinrich Ulrich developed a taxonomy that detailed two subtypes of the gay male Urning, namely Mannling and Weibling, for masculine and effeminate. Women who loved other women were Urningin. Dioning were heterosexual men. A little linguistic cross-pollination gives Urano-dioning for people who are attracted to both men and women. Virilisirt refers to a gay man who marries and lives as a heterosexual. The terms are clumsy but the scheme does acknowledge that people can be placed on a continuum rather than sorted into simple binary opposites. Other scientific terms of the age included invert and pervert.

The terms "lesbian" and "lesbianism" gradually replaced the older words "tribade" and "tribadism," which derive from the Greek and Latin for "to rub"—a concept also reflected in other languages, including Arabic, Hebrew, Swahili and Urdu.[9] By the 1920s in Japan the words for same-sex love between women were rezubian (from lesbian) and garuzon (from the French word garçon) referring to a woman considered to be more masculine. One effect of this adoption of language was to see lesbianism as a Western import.

Having come up with a name for people attracted to those of the same sex, Kertbeny also needed a name for those attracted to the opposite sex, and so, in 1888, heterosexuality was born. Heterosexuals had never been considered as a type that needed characterising or exemplifying. Heterosexuality is instead taken to be the standard or default against which everyone else is compared and judged.

Krafft-Ebing: Saving Civilisation

The Viennese psychiatrist Richard von Krafft-Ebing published his grandly titled *Psychopathia Sexualis* in 1886. Reprinted many times the book, an influential hit, helped to popularise the word homosexuality and created the template for discussing, examining and treating LGBTQ+ people.

This purportedly scientific study into human sexuality is organised into two parts: case studies and analysis. The case studies—which comprise the majority of the book—were largely provided by Krafft-Ebing's Berlin collaborator Albert Moll, who tapped his police connections for the subject matter. Moll's treatment of his material was designed to convey understanding and tolerance, but Krafft-Ebing turned these into something else entirely, and modern readers quickly see it for what it is: a work of obfuscation, condemnation and titillation with the more salacious passages rendered in Latin—further delighting publishers as this led to a boost to sales of Latin textbooks.

Around one-third of *Psychopathia Sexualis* is devoted to same-sex love and the majority of this is about men. He concludes, after much pseudo-medical labelling, that many homosexual men are "of a degenerative character"[10] who, because they are only interested in the pursuit of their own goals, represent a threat to moral order and society itself. Women are seen as asexual as was common at the time: "if she is normally developed mentally, and well bred, her sexual desire is small."[11]

Krafft-Ebing believed that his work would save civilised society. The book demonstrates a rigid, conservative hierarchy where men are superior to women, the refined upper classes are superior to the immoral masses, Christians are superior to Muslims, and the sexually normal are superior to the perverted. Krafft-Ebing's bestseller laid the groundwork for the further study of sex and greatly influenced the emerging profession of psychiatry.

Havelock Ellis: Shining a New Light

The influential British sexologist Havelock Ellis took a relatively liberal and open-minded approach to homosexuality, particularly when it came to men. His 1896 book on homosexuality proposed that it should not be considered a criminal offence.

John Addington Symonds, who was certainly bisexual or gay, helped Ellis write the book. From a very early age Symonds knew he was attracted to people of the same sex. He wrote about the ancient Greeks in an extended essay in 1876, making only ten copies which he gave to people in his network. In 1883 he privately distributed 50 copies of his next book, *A Problem in Greek Ethics*. Symonds gave these two publications to Ellis along with letters sent to him during the writing of the book. Symonds died of influenza before Ellis completed *Sexual Inversion* but was cited as the co-author.[12]

Ellis's principal motivation in writing the book was to shed light on sexuality and especially the topic of what was known as inversion. "Congenital inverts" were naturally attracted to those of the same sex who Ellis respected and admired despite being vilified by society. The widespread ignorance of the topic concerned him, and he was critical of physicians and theologians who treated homosexuality as an illness and a sin. He wanted to widen the range of experts who were able to observe and comment on sexual behaviour.[13]

Ellis accepted Krafft-Ebing's distinction between congenital and acquired inversion. Congenital inversion should be treated with sympathy and not considered unlawful. Ellis is less sympathetic towards acquired inversion as this involved individual choice. He also identifies "psychosexual hermaphroditism," a form of inversion where people are attracted to both sexes. This is something that he had doubts about, however, as he believed such individuals are really only interested in others of the same sex.

He also asserted that instances of same-sex love between women were less common than between men and more likely to happen in occupations that were predominantly female such as workers in large hotels, lacemaking and especially roles in the theatre. Such women were not that attractive to men, he declared, which was why they welcomed the attention of other females.

In addition to preferring male clothing some of the other signs of a female invert included being assertive, confident, wanting to have their voice heard, having opinions and refusing to accept the role of a woman in a mixed group. They were also more muscular, had a different type of larynx and preferred smoking to needlework. Furthermore, he was suspicious of the female emancipation movement because it promoted independence and which would inevitably, in his opinion, lead to a decline in marriage.

Sigmund Freud: Mind and Body

Freud showed that a patient's physical symptoms could be caused by psychological factors revealed by them talking about their experiences, dreams and thoughts. He sought to demonstrate that the unconscious mind has an impact on behaviour and decisions. Much of his work focused on child development, and he felt that many adult problems had their origins in early years' experience and he drew heavily on the work of earlier sexologists, including Krafft-Ebing and Ellis.

The triangular relationships between father, mother and child were the focus of his most influential theory about sexual development known as the Oedipus complex. The theory posits that the child goes through a number of stages in their sexual development, and when, they arrive at the phallic stage, the object of their attention turns to the mother, with whom they wish to have sex. The father is then seen as competition for the mother's attention.

Homosexuality couldn't be incorporated in the early versions of the theory. Up until the point that the child starts to turn their attention to the mother, the child will have identified with the father. The child would have behaved in quite feminine ways, seeking to get the father's attention, with the mother being seen as the rival. Hostility would have been directed towards her at this point. In normal sexual development this complex set of relationships is resolved by societal taboos regarding incest as well as the fear of castration. However, Freud theorised that some boys repressed their feelings for their mother and rather than being attracted to her they identified with her. This is unconscious, but as a consequence of identifying with a woman, it meant that they became attracted to men.

Freud felt that heterosexual relationships reflected full development whereas same-sex attraction stemmed from repressed ideas and emotions. In pursuing other men, people were actually seeking to avoid other women in order not to be unfaithful to their mother.[14]

Freud identified three types of inverts: "absolute inverts," who are attracted to people of the same sex exclusively; "amphigenic inverts, that is psychosexual hermaphrodites" where the sexual object could be someone of either sex; and "contingent inverts," where someone is attracted to someone of the same sex due to the circumstances they find themselves in.[15] He believed that inversion was variable within groups and individuals.[16] Significantly, he rejected the idea of inversion being a form of degeneracy, making a distinction between inverts and perverts, the latter including fetishists.[17]

His attitudes are further clarified in a letter Freud wrote to "A Grateful Mother":

I gather from your letter that your son is a homosexual. I am most impressed by the fact that you do not mention this term yourself in your information about him. May I question you why you avoid it? Homosexuality is assuredly no advantage, but it is nothing to be ashamed of, no vice, no degradation, it cannot be classified as an illness; we consider it to be a variation of the sexual function produced by a certain arrest of sexual development. Many highly respectable individuals of ancient and modern times have been homosexuals, several of the greatest men among them (Plato, Michelangelo, Leonardo da Vinci, etc.) It is a great injustice to persecute homosexuality as a crime and cruelty too. If you do not believe me, read the books of Havelock Ellis.[18]

The Oedipus complex however had no place for bisexuality. Freud therefore theorised bisexuality as a form of arrested development—a regression to a more primitive state.[19]

More widely, bisexuality was seen as a primitive stage in the evolution of civilised societies: "it was relegated to a place and time outside culture, to the sphere of pre-history. In this way bisexuality was a necessary blindspot for sexological discourse."[20] It is part of Freud's legacy that bisexuality was not considered to be a legitimate identity for many decades to come. As Steven Angelides puts it:

In the three decades following Freud's death the concept of bisexuality was almost unilaterally repudiated as a scientific falsehood within the domains of psychoanalysis and psychiatry. In addition to this, homosexuality was refigured as a form of pathology.[21]

Magnus Hirschfeld: Data and Defiance

Another major figure in the development of our understanding of human sexual behaviour was the German physician and psychiatrist Magnus Hirschfeld, who was also gay. He was enormously active in a wide range of areas connected with human sexuality—even supplying a foreword to a 1929 guide to 14 clubs and bars, Berlins Lesbiche Frauen (Berlin's Lesbian Women). He founded the Scientific-Humanitarian Committee (WhK), which campaigned for equal rights for homosexuals. He also established the Institute for Sexual Science, which quickly gained an international reputation for the thoroughness of its work and its preparedness to provide information on an aspect of human behaviour which others preferred to remain ignorant about while choosing to stigmatise.[22] He created the first publication for the scientific study of homosexuality and homosexual behaviour, the *Jahrbuch für sexuelle Zwischenstufen* (Yearbook for Sexual In-between), which ran from 1899 to 1923.

Hirschfeld was determined to gather data and statistics rather than rely on case studies. In 1903 he carried out a survey of 3,000 students and 5,000 factory workers which essentially consisted of one question: Are you attracted to women and men or both? Of the students, 6% were not heterosexual (4.5% were bisexual, 1.5% homosexual). Of the factory workers, the total who were not heterosexual was lower at 4.34% (3.19% bisexual and 1.15%

homosexual). His overall conclusion was that 2.2% of men in Germany were homosexual, with a further 3.2% identifying as bisexual. He also suggested that people felt unable to express themselves fully and were hiding their sexual orientation from the rest of society. Hirschfeld was charged with indecency on publication, but the case was thrown out.[23]

A constant thorn in the side of the authorities, Hirschfeld co-wrote and acted as a doctor in the first movie to explore gay love, *Anders als Die Andern* (Different from the Others, 1919). The intertitles convey clearly the film's intentions to view same-sex love as natural and as such not to see it as a crime.

As the doctor, he says:

> because he is a homosexual, he is not to blame for his orientation. It is not wrong, nor should it be a crime. Indeed, it is not even an illness, merely a variation, and one that Is common to all of nature . . . Love for one's own sex can be just as pure and noble as that for the opposite sex.

The film was also sending a message to those who had been made to feel different, ashamed and guilty: "Only ignorance or bigotry can condemn those who feel differently. Don't despair! As a homosexual, you can still make valuable contributions to humanity." Only fragments of the film survived the Nazi period.[24] Hirschfeld was shot at while giving a lecture in Vienna in 1923 and several members of the audience were wounded.[25]

He continued his work undeterred. The WhK had a peak membership of 500 with branches in over 200 cities in Germany, Austria and the Netherlands.[26] His clinics provided counselling on sexual matters as well as marriage guidance and advice on contraception. Meanwhile his campaigning committee battled to decriminalise homosexuality in Germany, fighting the rising tide of National Socialism. After Hitler's appointment as chancellor in 1933, Nazis stormed Hirschfeld's Scientific-Humanitarian Committee and his scientific institute, taking his survey data to help identify, track down and imprison the homosexual respondents.[27] Hirschfeld went into exile soon afterwards and died in 1935.

Laws and Wars

Counterbalancing movements were also taking place. Decriminalisation of homosexuality occurred in Denmark in 1933, Iceland in 1940, Switzerland in 1942 and Sweden in 1944. Some countries such as Poland, Russia and Czechoslovakia, which had previously been more open, passed repressive laws in the years before and during the Second World War. France and Germany outlawed homosexuality under Nazi control but repealed these measures in 1956 (Germany) and 1982 (France). The Sexual Offences Against the Person Act decriminalised homosexual acts in private between consenting adults aged 21 or over in England and Wales in 1967, with decriminalisation in Scotland in 1981 and in Northern Ireland in 1982. However, social attitudes did not keep pace with legislative change.[28]

Sexologists' research methods moved on from case studies and surveys to much more intrusive methods. The Committee for the Study of Sex Variants was set up in New York in 1935 to "undertake, support and promote investigations and scientific research, touching upon and embracing the clinical, psychological and sociological aspects of variations from normal sexual behaviour . . . through laboratory, research and clinical study."[29] The researchers were keen to show that it was possible to identify sexual deviance by physical characteristics and differences.

Participants in the study were interviewed about their family background and undertook psychiatric assessments. Women were subjected to

> incredibly intrusive examinations, which involved measuring the clitoris with a ruler and a vagina with fingers, as well as tracing the vulva on a glass plate, [by which means] the researchers sought to detect such signs as a large vulva, erectile clitoris, insensitive hymen, and small uterus.[30]

"Scientific" analyses such as these were accorded as much respect as they are derided now.

The US Army and Navy developed their own pseudoscientific and denigratory labels. Taking their lead from Krafft-Ebing, homosexuality was a "constitutional, psychopathic state" and homosexuals "sexual psychopaths."[31] The powerful combination of military officials and psychiatrists ensured that homosexuality was viewed as a mental illness. From 1941, anybody found to be a homosexual could be discharged dishonourably from the military. At the Fort Leavenworth base in Kansas, anyone caught having sex with another man was forced to wear a yellow D on their backs, which stood for "Degenerate"—an uncanny echo of the Nazis' system for labelling the minorities they persecuted and murdered. The US Navy even designated a special naval prison in Portsmouth, New Hampshire, for "moral perverts."[32]

Conversely, the war also provided opportunities. People moved from villages and towns to cities and were better able to meet with like-minded people.[33] Many of them stayed where they had been stationed and so their wartime networks continued after the war. Gay bars started to appear along with the first publications promoted specifically among gay communities. Organisations formed calling for equal rights, one of the first being the Mattachine Society, established in 1948 by Henry Hay in America and taking inspiration from masked protest dances by the poor in medieval France. For Hay this was an ideal metaphor for the situation that gay people faced in America: having to remain anonymous to avoid being stigmatised and discriminated against.[34]

Psychiatry Supplants Religion

One of the biggest threats posed to the LGBTQ+ community after the war came from psychiatrists and psychoanalysts. Despite being young professions, their influence had grown considerably and they began to accept the prevailing attitudes and norms of society, including increasingly conservative approaches to sexual orientation. These professionals felt they had unique insights into human sexuality, and they would promote, strengthen and defend their specialism vigorously if attacked.

Freud may have seen homosexuality as an inversion, but for his followers, it was a perversion.

This process started almost as soon as Freud passed away in 1940. Psychoanalyst Sandor Rado exemplified the professional need to be certain, confident and right. For him the causes of homosexuality were straightforward: it was the parents' fault. By not providing the right sex education, parents left children with feelings of anger, fear and anguish, which lead to neuroticism, the basis of homosexuality.[35] For Sandor, as with Freud, there was only heterosexuality and homosexuality—a distinction that also represented good and evil as well as health and sickness. Bisexuality did not and could not exist.

Psychiatry effectively took over the role of the church, using the discipline's learning to reinforce the prejudices of society rather than challenging them. Their hold over how homosexuality should be viewed became even stronger with the arrival of the Diagnostic and Statistical Manual of Mental Disorders (DSM) in 1952. Based on classification work carried out by the US military during the war, the first edition of the DSM described homosexuality as a "sociopathic personality disturbance" alongside transvestitism, paedophilia, fetishism and sexual sadism.[36]

A generous interpretation of the psychiatrists' motivation would be that they wanted to free homosexuality from association with immorality. However, the DSM plainly codified homosexuality as abnormal. Post-war psychiatry monopolised sexuality: psychiatrists could identify homosexuals, and they could cure them too. They were the new, high-status saviours of civilised society. And business boomed.[37]

Alfred Kinsey: Science Fights Back

Despite the ascendancy of psychiatry, the field increasingly came under attack from scientists. The work of Alfred Kinsey was key here. Formerly the world's expert on gall wasps, Kinsey took a sideways move into the field of human sexuality in 1938 when the University of Indiana asked him to help design a course on marital relationships. Kinsey soon "discovered that scientific understanding of human sexual behaviour was more poorly established than the understanding of almost any other function of the human body."[38] He found that much of the existing research was based on small samples, biased in its methodology and subjective in its conclusions. As an experienced scientific researcher, he did what came most naturally to him: he carefully designed studies to yield objective, statistical data instead of opinion and rhetoric.

His overall sample population was huge: 5,300 white males and 5,940 white females. The method was in-depth personal interviews lasting between 90 and 120 minutes. The two major publications produced from this research, *Sexual Behaviour in the Human Male* (1948) and *Sexual Behaviour in the Human Female* (1953), caused a sensation. The key findings include:

- 37% of males and 13% of females had at least some overt homosexual experience to orgasm.
- 10% of males were more or less exclusively homosexual and 8% of males were exclusively homosexual for at least three years between the ages of 16 and 55. For females, Kinsey reported a range of 2–6% for more or less exclusively homosexual experiences/responses.
- 4% of males and 1–3% of females had been exclusively homosexual after the onset of adolescence up to the time of the interview.[39]

Re-analysis in 1972 showed that between 25% and 33% of white males had had an overt homosexual experience. Overall, it was estimated that about 4% of white males were predominantly or exclusively homosexual and for women the figure was between 1% and 2%.[40]

Kinsey poured scorn on the heterosexual-homosexual binary:

The world is not to be divided into sheep and goats. Not all things are black nor all things white. It is a fundamental of taxonomy that nature rarely deals with discrete categories. Only the human mind invents categories and tries to force facts into separated pigeon-holes. The living world is a continuum in each and every one of its aspects. The sooner we learn this concerning human sexual behaviour the sooner we shall reach a sound understanding of the realities of sex.[41]

He created a sexuality rating scale, now commonly referred to as the Kinsey scale, which goes from 0 (exclusively heterosexual) to 6 (exclusively homosexual). Sexuality therefore had a considerable degree of fluidity, and it wasn't possible to put people into absolute categories as psychiatry was so insistent on doing. Same-sex attraction couldn't be considered to be a form of abnormality: "it is difficult to maintain the view that psychosexual reactions between individuals of the same sex are rare and therefore abnormal or unnatural, or that they constitute within themselves evidence of neurosis or even psychosis."[42] He was critical of the legal structures regarding sexuality and the hypocrisy of those in authority who condemned in others behaviour that they themselves engaged in. There can be little doubt that he also took pleasure in debunking the theories of psychoanalysts.

The reaction from the psychiatric and psychoanalytic community was furious and swift. His work was dismissed on every level possible—from the selection of the sample to the sheer ridiculousness of his conclusions. Edmund Bergler claimed access to a higher form of truth on the part of psychoanalysis, maintaining that psychoanalytic theory was based on genetics, which was superior to any quantitative analysis. He also offered the thought that "there are no happy homosexuals; and there would not be, even if the outer world left them in peace."[43]

Evelyn Hooker and the Attack on Psychoanalysis

Psychologist Evelyn Hooker's ground-breaking research study of 1957 revealed how subjective and judgmental psychoanalysts were in their assessments of gay people. She set out to "investigate the adjustment of the homosexual"[44] by examining subjects who did not come from the usual medical or criminal sources and comparing them to a matched group of heterosexual men.

Each participant completed three tests commonly used by the psychoanalytic community: the Rorschach Test, the Thematic Apperception Test (TAT) and the Make-A-Picture-Story Test (MAPS). The Rorschach Test asks people to say what they see in abstract, "inkblot" images. The TAT asks respondents to describe what they think is going on in a series of pictures, including the emotions of the characters. MAPS involves participants using a backdrop and cardboard cut-out figures to tell a story.

The tests were interpreted by psychoanalysts. Hooker then looked to see whether one group of people could be distinguished from the other based on the psychoanalysts' interpretations. The Rorschach Tests were far from illuminating, with one rater admitting, "I just have to guess."[45]

It was easier for the psychoanalysts to identify some homosexual participants from the narrative elements in the (separately analysed) TAT and MAPS tests. But in overall terms, they could not distinguish between the homosexual and heterosexual participants. Some of the analysts even showed signs of questioning their own views about what "normal" looks like. Where a participant had described a same-sex relationship, the psychoanalysts' training would have led them to categorise the individual as having some form of psychiatric illness. One of the analysts, reflecting on the profile he had written becomes concerned about the categorisation of homosexuality as a form of illness: "this record is schizophrenic like I am an aviator. If you want proof that a homosexual can be normal, this record does it" (p27). Tellingly however, profiles which revealed them to be well-adjusted individuals were more likely to be judged as heterosexual for example:

He is able to love and to dislike. He is a good father and husband and would be a steady employee. I could see him as having a better than average job. He would not be a creative or imaginative person. I don't mean a Babbitt ["a materialistic, complacent, and conformist businessman" OED] but he would not take the risk of looking deeply. He is a middle-of-the-roader. This is as clean a record as I think I have seen. I don't think he has strong dependency needs. He is comfortable, and in that sense, he is strong. Imply that this is a heterosexual record specifically.

(p28)

But once the sexual orientations of the same participants were revealed, the analysts' judgements became much more certain and categorical:

What is here is indecision and a schizoid feeling. So this is not in any sense, a superior personality. There is some withdrawal and some aridity. This is not an outgoing, warm, decisive person. It is a constricted, somewhat egocentric, somewhat schizoid, perturbed, a little guilty fellow.

(p28)

Once an indication is given of the participants' sexuality, the descriptions couldn't be more different.

By classifying the descriptions that the psychoanalysts provided we can see clearly the mental models they were using and how they viewed homosexuals, as shown in Table 3.1.

Hooker's conclusion was that "homosexuality as a clinical entity does not exist. Its forms are as varied as those of heterosexuality." Furthermore "homosexuality may be a deviation in sexual pattern, which is within the normal range, psychologically."[46] Hooker's intention in conducting the research was to show that homosexual men were as well adjusted as heterosexual men. Her tone towards psychiatry and its practitioners was respectful and grateful. She acknowledges the time and effort it had taken for the professionals to carry out the huge amount of work required for the analysis. It was not her intention, we believe, to expose the failings of the discipline.

Despite Hooker's respectful attitude towards the psychoanalytic profession, it was clear that her research identified the biases against homosexuality within the profession.

Table 3.1 Adjectives used to describe heterosexuals, and homosexuals by psychoanalysts

Heterosexual	Homosexual
Clean	Schizophrenic
Strong	Withdrawn
Low dependency needs	Perturbed
Comfortable	Guilty
Good	Dependency needs
Not creative	Constricted
Lacks imagination	Egocentric
Steady	Indecisive
Able to love	Lacks warmth
	Not outgoing

Clellan S. Ford and Frank A. Beach—whose backgrounds were in anthropology and ethology—in 1951 published their book *Patterns of Sexual Behaviour*. Their research aims remain incredibly ambitious even today. They examined sexual behaviour across many cultures, identifying similarities as well as differences. They ascribed the differences to the rules and norms of each society—that is, what was considered acceptable or not. They also extended their reach across species.

Ford and Beach found that same-sex attraction was absolutely normal and natural. Where societies viewed it otherwise, this was due to their attitudes. The authors looked at 76 societies and found that homosexual activity was condemned in 28 (36%) of them. These condemning societies denied that such behaviour occurred—and maintained that if it did, it was very rare. "The penalties range from the lighter sanction of ridicule to the severe threat of death."[47] They also found societies where "males are singled out as peculiar if they do not indulge in these homosexual activities."[48]

They could only obtain data on same-sex relationships between women from 17 societies. Contradictory explanations were given about why women would be attracted to one another. Among the Dahomeans homosexuality in women was seen as the reason for frigidity in their marriages, but for Haitians, it was felt that a woman would seek to have a woman as a partner because she couldn't satisfy her man. The research also identified a few people in each society who could be termed intersex. From their comparison of humans to other animals, they concluded that "inversion of the sexual rule is common among animals of several species other than homo sapiens, and it is particularly frequent in infrahuman primates."[49]

Thomas Szasz went further with his furious attack on psychiatry and psychoanalysis in *The Myth of Mental Illness*. The categorisation of homosexuality as an illness was for him an example of the power grab that psychiatry was undertaking:

This reclassification of non-illnesses as illnesses has, of course, been of special value to physicians and to psychiatry as a profession and social institution. The prestige and power of psychiatrists have been inflated by defining ever more phenomena as falling within the purview of their discipline. . . . It is difficult to see why we should permit, much less encourage, such expansionism in a profession and so-called science.[50]

One psychotherapist who insisted that homosexuality was a form of abnormality was Charles Socarides. His starting point was mother fixations and narcissism from which he riffed that homosexuality

can be seen as a resolution of the separation from the mother by running away from all women. In fantasies and actions, in reality, in the compulsive hunting for partners, the homosexual is unconsciously searching for the lost objects, seeking to find the narcissistic relationships he once experienced in the mother–child symbiosis.[51]

To read Socarides today sparks awe at the scope of his imagination and the ridiculousness of his assertions:

> the penis of the male partner becomes the substitute for the mother's breast. . . . [H]e divests himself of Oedipal guilt, by demonstrating to her [the mother] that he could have no possible interest in other females. He is interested only in men. . . . Furthermore, he is protecting the mother against the onslaught of other men's penises, allowing penetration into himself instead.[52]

Socarides continued to claim well into his career that he could cure homosexuals. When asked by *Esquire* magazine about his gay son, who went on to be an adviser on gay and lesbian rights in the Clinton administration, he said that he was "very proud" of Richard and also that "people shouldn't be ashamed of homosexuality."[53] It's startling to realise that Socarides had no idea of the impact his work was having on the stigmatisation of LGBTQ+ people.

The year 1968 saw the publication of DSM-II with homosexuality now reclassified among the personality disorders. The world had changed between the two editions of the DSM and change accelerated thereafter. Since the police raid on the Stonewall Inn in New York in 1969, which led to widespread disorder, members of the LGBTQ+ community had become more vocal and visible in demanding equal rights. Demonstrations took place at the American Psychiatric Association's 1970 meeting in San Francisco over the continued inclusion of homosexuality in the DSM. In theory there could be no LGBTQ+ people within the APA as this was considered to be a disorder. At the 1971 conference gay rights activists were invited to be part of a panel discussion entitled "Gay is Good." They made it clear that the problems of stigmatisation and discrimination that members of the community faced were in large part due to the way psychiatry viewed them. The 1972 gathering saw another panel discussion with one panellist wearing a huge mask, his voice distorted by a microphone. Referred to as Dr H Anonymous, he was in fact a gay psychiatrist who, fearing for his career, had agreed to participate only if he was heavily disguised. Psychiatry, it was argued, had taken on board the attitudes of society rather than using their professional expertise to challenge them. APA vice president Judd Marmor argued that "psychiatry is prejudiced."[54]

These events led to a change in psychiatric language in 1973 when homosexuality became "sexual orientation disturbance." In other words, same-sex attraction was no longer the issue: the focus of psychiatric therapy should be the emotional disturbance that such an attraction might cause, which could be a consequence of societal attitudes and fear of being stigmatised.

Fifty-eight per cent of the APA's membership approved of the change in the nomenclature. The definition and description of homosexuality continued to change, becoming "ego-dystonic homosexuality"—a type of psychosexual disorder—in the 1980 edition of the DSM before finally being removed in 2013.

Key Points

The templates for attitudes that prevailed in the 20th century and are still present today were laid down by the early sexologists. The work that was most influential in our view was *Psychopathia Sexualis* published in 1886 written by Richard von Krafft-Ebing. This is established homosexuality as a deviance so abhorrent that society itself was at risk. Homosexuals were degenerates whose behaviour needed to be curtailed to prevent the spread of their infection. He described treatments that could cure the individuals and so leading to the conversion therapies that are still in use today. (Interestingly his book also inadvertently demonstrates the in effectiveness of these treatments).

The popularity of the book globally meant that societies that previously had an accepting attitude towards diversity of sexuality now became much more hostile. Colonisation, imperialism and racist hierarchies, which saw anyone who was not white as being inferior, also meant that legislation criminalising people who loved those of the same sex was implanted into territories that had never seen such behaviour as a problem before.

The new disciplines of psychotherapy and psychiatry saw an opportunity to increase their influence by explaining how loving someone of the same sex was wrong but also could be explained. This also meant that these professions could also provide suitable treatment for those affected. At the same time, however, there was also considerable resistance to these ideas and methodologies. Alfred Kinsey was one of a number of influential scientists who conducted robust research which showed the irrationality and the prejudice of those who viewed anything other than heterosexuality as an abnormality.

It's also important to remember that it was only relatively recently that homosexuality stopped being viewed as a form of illness.

Notes

1 Feray, J.-C., Herzer, M., & Peppel, G.W. (1990). Homosexual Studies and Politics in the 19th Century: Karl Maria Kertbeny. *Journal of Homosexuality*, 19(1), pp. 23–48.
2 Feray, J.-C., Herzer, M., & Peppel, G.W. (1990). Homosexual Studies and Politics in the 19th Century: Karl Maria Kertbeny. *Journal of Homosexuality*, 19(1), pp. 23–48.
3 Smedley', A., & Smedley, B. D. (2005). Race as biology is fiction, racism, as a social problem is real: Anthropological and historical perspectives on the social construction of race. *Am. Psychol.*, 60, 16–26.

4 Spencer, C. (1996). *Homosexuality: A History.* United Kingdom: Fourth Estate.
5 Angelides, S. (2001). *A History of Bisexuality* (1st ed.). The University of Chicago Press, p. 24.
6 Spencer, C. (1996). *Homosexuality: A History.* United Kingdom: Fourth Estate.
7 Rupp, L.J. (2009). *Sapphistries: A Global History of Love between Women.* NYU Press.
8 Spencer, C. (1996). *Homosexuality: A History.* United Kingdom: Fourth Estate, p. 10.
9 Rupp, L.J. (2009). *Sapphistries: A Global History of Love between Women.* NYU Press.
10 von Krafft-Ebing, R. (2012). *Psychopathia Sexualis* (Classic Reprint). Forgotten Books, p. 226.
11 von Krafft-Ebing, R. (2012). *Psychopathia Sexualis* (Classic Reprint). Forgotten Books, p. 13.
12 Mondimore, F.M. (1996). *A Natural History of Homosexuality.* Johns Hopkins University Press.
13 Ellis, H. (1915). *Sexual Inversion* (3rd ed.). London: Wilson and Macmillan.
14 Hertzmann, L., & Newbigin, J. (2023). *Psychoanalysis and Homosexuality: A Contemporary Introduction.* Routledge.
15 Freud, S., & Strachey, J. (1986). Three Essays on the Theory of Sexuality: I: The Sexual Aberrations. In: *Essential Papers on Object Relations*, pp. 5–39.
16 Freud, S., & Strachey, J. (1986). Three Essays on the Theory of Sexuality: I: The Sexual Aberrations. In: *Essential Papers on Object Relations*, pp. 5–39.
17 Hertzmann, L., & Newbigin, J. (2023). *Psychoanalysis and Homosexuality: A Contemporary Introduction.* Routledge.
    Freud, S., & Strachey, J. (1986). Three Essays on the Theory of Sexuality: I: The Sexual Aberrations. In: *Essential Papers on Object Relations*, pp. 5–39.
18 Freud, S. A Letter from Freud. *The American Journal of Psychiatry*, 107(10), pp. 786–787.
19 Angelides, S. (2001). *A History of Bisexuality* (1st ed.). The University of Chicago Press, p. 63.
20 Angelides, S. (2001). *A History of Bisexuality* (1st ed.). The University of Chicago Press, p. 48.
21 Angelides, S. (2001). *A History of Bisexuality* (1st ed.). The University of Chicago Press, p. 72.
22 Mondimore, F.M. (1996). *A Natural History of Homosexuality.* Johns Hopkins University Press.
23 Ledain, D. (2019). *This Forbidden Fruit: Male Homosexuality, a Culture and History Guide.*
24 Ledain, D. (2019). *This Forbidden Fruit: Male Homosexuality, a Culture and History Guide.*
25 Mondimore, F.M. (1996). *A Natural History of Homosexuality.* Johns Hopkins University Press.
26 Ledain, D. (2019). *This Forbidden Fruit: Male Homosexuality, a Culture and History Guide.*
27 Mondimore, F.M. (1996). *A Natural History of Homosexuality.* Johns Hopkins University Press.
28 *Psychoanalysis and Homosexuality (Introductions to Contemporary Psychoanalysis).* Routledge.
29 Terry, J. (1999). *An American Obsession.* University of Chicago Press, p. 178.
30 Rupp, L.J. (2009). *Sapphistries: A Global History of Love between Women.* NYU Press.
31 Spencer, C. (1996). *Homosexuality: A History.* United Kingdom: Fourth Estate, pp. 347–348.

32 Spencer, C. (1996). *Homosexuality: A History*. United Kingdom: Fourth Estate, p. 349.
33 Mondimore, F.M. (1996). *A Natural History of Homosexuality*. Johns Hopkins University Press.
34 Spencer, C. (1996). *Homosexuality: A History*. United Kingdom: Fourth Estate.
35 Tontonoz, M.(2017). Sandor Rado, American Psychoanalysis, and the Question of Bisexuality. *History of Psychology*, 20(3), p. 263.
36 https://ajp.psychiatryonline.org/doi/full/10.1176/appi.ajp-rj.2022.180103
37 Tontonoz, M. (2017). Sandor Rado, American Psychoanalysis, and the Question of Bisexuality. *History of Psychology*, 20(3), p. 263.
38 Kinsey, A.C., Pomeroy, W.B., Martin, C.E., & Gebhard, P.H. (1953). *Sexual Behavior in the Human Female*. Saunders, p. 5.
39 https://kinseyinstitute.org/research/publications/historical-report-diversity-of-sexual-orientation.php#:~:text=The%201948%20and%20 1953%20Studies%20of%20Alfred%20Kinsey&text=26).,almost%20all%20 of%20the%20data
40 https://kinseyinstitute.org/research/publications/historical-report-diversity-of-sexual-orientation.php#:~:text=The%201948%20and%20 1953%20Studies%20of%20Alfred%20Kinsey&text=26).,almost%20all%20 of%20the%20data
41 Kinsey, A.C., Pomeroy, W.B., & Martin, C.E. (1948). *Sexual Behavior in the Human Male*. Saunders, p. 639.
42 Kinsey, A.C., Pomeroy, W.B., & Martin, C.E. (1948). *Sexual Behavior in the Human Male*. Saunders, p. 659.
43 Bergler, E. (1948). The Myth of a New National Disease. *Psychiatric Quarterly*, 22, pp. 66–88. https://doi.org/10.1007/BF01572406
44 Hooker, E. (1957). The Adjustment of the Male Overt Homosexual. *Journal of Projective Techniques*, 21, pp. 18–31.
45 Hooker, E. (1957). The Adjustment of the Male Overt Homosexual. *Journal of Projective Techniques*, 21, pp. 18–31.
46 Hooker, E. (1957). The Adjustment of the Male Overt Homosexual. *Journal of Projective Techniques*, 21, pp. 18–31.
47 Ford, C.S., & Beach, F.A. (1951). *Patterns of Sexual Behavior*. Harper and Paul B. Hoeber, p. 129.
48 Ford, C.S., & Beach, F.A. (1951). *Patterns of Sexual Behavior*. Harper and Paul B. Hoeber, p. 132.
49 Ford, C.S., & Beach, F.A. (1951). *Patterns of Sexual Behavior*. Harper and Paul B. Hoeber, p. 134.
50 Szasz, T. *The Myth of Mental Illness* (Kindle ed.). HarperCollins, p. 28, p. 39.
51 Socarides, C.W. (1968). A Provisional Theory of Aetiology in Male Homosexuality-a Case of Preoedipal Origin. *The International Journal of Psycho-Analysis*, 49, p. 27. https://go.openathens.net/redirector/leeds.ac.uk?url=www.proquest.com/scholarly-journals/provisional-theory-aetiology-male-homosexuality/docview/1298184610/se-2
52 Socarides, C.W. (1968). A Provisional Theory of Aetiology in Male Homosexuality-a Case of Preoedipal Origin. *The International Journal of Psycho-Analysis*, 49, p. 27, p. 29.
53 Walls, J. (1997, 05). Father Knows What? *Esquire*, 127, p. 23. https://go.openathens.net/redirector/leeds.ac.uk?url=www.proquest.com/magazines/father-knows-what/docview/210277512/se-2
54 https://ajp.psychiatryonline.org/doi/full/10.1176/appi.ajp-rj.2022.180103

Part II

Part LGBTQ+ in the Workplace

Chapter 4

Stereotypes and Stigma at Work

As we have seen, attitudes and stigma towards the LGBTQ+ community have been shaped by religion, law and science over generations. The modern organisational setting is a product of the same forces but the experiences of LGBTQ+ employees have been further affected by relatively recent events and phenomena. This chapter explores the impact of societal attitudes on LGBTQ+ in the workplace and the strategies people use to counteract the stigma they experience.

Stereotypes

Societal attitudes towards groups of people are driven by stereotypes held about them. Stereotypes are the oversimplified, preconceived notions or beliefs that we hold about certain groups of people. They are generalisations based on limited information. They also often stem from our personal experiences, such as interactions with individuals within a group, the beliefs of our family and cultural influences such as representations in the media. Research shows that once a representation of a group is formed in our mind, it is extremely resistant to change, even when confronted with opposing evidence.[1]

Stereotypes are both prescriptive—generating expectations about how people should behave and what they should do—and descriptive—generating expectations about how people are.[2,3,4]

Many examples of stereotypes are widely held and well reported in the research: for example, women are stereotyped as being compassionate, kind and helpful while men are stereotyped as being strong, ambitious and independent.[5] Stereotypes are typically connected, meaning that we link stereotypes about one aspect of a person, such as their appearance, with assumptions about their interests or abilities. For example, masculinity will evoke assumptions about leadership competence as this is a stereotypically male trait.[6]

Our fixed assumptions about groups of individuals also translate into ideas about different types of work. For instance, we tend to associate professions

DOI: 10.4324/9781003489580-6

such as surgeon, basketball player and carpenter with men, and professions such as nurse, childcare assistant and beautician with women. The impact is that when an individual doesn't fit the "image" for a particular role, they may be overlooked despite being suitably qualified and having adequate experience.

One aspect of this phenomenon which has been researched extensively is the similarity in stereotypes of men and leaders, known as the Think Manager, Think Male paradigm.[7] This is a global phenomenon evident in both men and women that ultimately leads to women being deemed less effective leaders[8] and restricts their access to promotions and development opportunities. Extending this paradigm, the stereotypes of women that we hold are closely associated with follower-behaviours, suggesting that when we think follower, we think female.[9] This effect contributes to the glass ceiling, the invisible barrier faced by women seeking leadership roles.

Conformity to traditional stereotypes disadvantages women in their pursuit of leadership roles for not fitting the stereotypical ideal of leaders. But what about women who don't conform to traditional stereotypes of women? Women who are considered to possess the necessary traits for leadership, such as competitiveness, ambition, competence and assertiveness, face challenges, barriers and repercussions from others[10] known collectively as the backlash effect.[11] And this is not just about women: men who assume traditionally feminine roles are judged to be ineffectual and less respected than women.[12]

Stereotypes about a group don't just impact the way that others interact with them. They also influence the way individuals see themselves and impact their behaviours and performance.

Almost anything that reminds an individual about their stigmatised identity can impact behaviour. In fact, simply ticking a box to indicate your gender or your race can impact test scores through the invoking of stereotypes regarding ability and performance.[13] This phenomenon, known as stereotype threat, arises when activated stereotypes burden the individual with distracting thoughts and self-conscious beliefs about their behavior and performance. This mental preoccupation, coupled with anxiety over conforming to the stereotype, imposes a cognitive load that diminishes the individual's ability to focus and perform effectively on the task at hand.

In one research study women were exposed either to gender-neutral TV adverts or to gender-stereotypic ones designed to elicit and reinforce common stereotypes of women. They were then asked to indicate whether they would prefer to assume a leadership or problem-solving role in a subsequent task. While the women exposed to the neutral adverts did not indicate a preference either way, exposure to the gender-stereotypic advert led women to express less interest in the leadership role.[14]

Anticipation of judgement or negative treatment by others undermines individuals' performance and aspirations, presenting a perceived career barrier.[15] The belief that someone will meet career barriers further down the line can lead to decisions not to pursue development opportunities. This effect has been identified as a contributing factor in the underrepresentation within specific occupations,[16] and exacerbates the "leaky pipeline" seen in some industries such as those related to science, technology, engineering and mathematics (STEM).[17] Unfortunately we don't need to wait until an individual's career is under way to see the impact of this. Young children become aware of gendered and professional stereotypes very early on, contributing to different interests in boys and girls.[18]

LGBTQ+ Stereotypes

The LGBTQ+ acronym encapsulates many different identities. Early research in this domain focused on the negative perceptions and beliefs regarding gay men, for example, that they are "mentally ill"[19] or "abnormal."[20] Several measures of attitudes towards gay men and lesbians were developed during this time, including the Attitudes Toward Lesbians and Gay Men Scale,[21] the Homosexuality Attitudes Scale[22] and the Modern Homonegativity Scale.[23] Research using such measures revealed that negative beliefs regarding lesbians and gay men were pervasive and that they played a significant role in the incidence of homophobia, discrimination and prejudice.

These measures rely on self-reporting of explicit attitudes. More recent research has gone beyond self-reported attitudes to uncover the subtler, less conscious biases that individuals may hold. One prominent method for studying these implicit biases is the Implicit Association Test (IAT), which has been adapted to examine attitudes related to sexuality. This approach has shed light on the implicit stereotypes that people may hold regarding non-heterosexual individuals, offering insights beyond the surface-level perceptions of positivity or negativity. By exploring both explicit and implicit attitudes, we gain a more comprehensive understanding of how beliefs and biases impact the LGBTQ+ community and contribute to the broader issues of homophobia, discrimination and prejudice.

An examination of IAT data from its inception in 2002 to 2006 revealed that 68% of participants displayed an implicit preference for heterosexual individuals over gay individuals.[24] Similar research conducted between 2006 and 2012 confirmed the persistence of this preference.[25] However, just as with the earlier scales and assessments, this line of research primarily addresses perceptions of favourability or unfavourability towards lesbians and gay men rather than delving into the precise nature of these attitudes and stereotypes.

Beyond broad perceptions of negativity and unfavourability the content of stereotypes about LGBTQ+ individuals differ depending on specific identity. That said, stereotypes of gay men and lesbians share one thing in common: they are based on the idea of gender non-conformity The stereotypes commonly associated with them tend to be the opposite of those held about their heterosexual equivalents. Where heterosexual women are stereotyped as feminine and heterosexual men are assumed to be masculine, lesbians are believed to be more masculine and gay men are assumed to be highly feminine.[26] This phenomenon was initially described in 1987 as the Implicit Inversion Theory,[27] and these beliefs have since been found to be held by lesbians and gay men themselves as well as heterosexuals.[28] The effect extends to the associated assumptions we make about behaviours and skills: gay men are rated as stereotypically warmer than lesbians and lesbians are rated as more competent than gay men.[29] The close coupling of these assumptions means that once a person believes somebody is gay or a lesbian, their expectations of that person's ability, personality and preferences will also be gender-atypical.[30]

Researchers use trait assignment tasks to investigate the content of these stereotypes. These tasks involve asking participants to assign a list of masculine and feminine traits to heterosexual women and men, and to lesbians and gay men. Over time the stereotypes of gay men and lesbians have become more androgynous, that is, possessing both masculine and feminine traits.[31] A slightly different methodology which asks participants to freely identify traits and stereotypes instead of picking items from a list has found that beliefs about lesbians and gay men are indeed multifaceted.[32] This approach has allowed for further refinement of stereotype content and the emergence of different types of identity such as "hyper-masculine" gay men[33] and "lipstick" lesbians.[34] Although sub-categories here are still emerging, these findings lend support to the general categorisation of gay men as stereotypically effeminate and lesbians as typically masculine.

These stereotypes have become widely accepted and are evident in the language we use and in media representations of lesbians and gay men. The strength and wide acceptance of these stereotypes, titled homonormativity by Lisa Duggan,[35] have led to the idea that there is an "acceptable" way to be gay, which is to behave in sync with the existing norms.

The conflation of gay male identity with traits associated with heterosexual women has also led to the belief that, like women, gay men are unsuitable for leadership roles. In fact, the prototypes for successful managers are more closely associated with heterosexual male and female managers than to stereotypes of gay male managers.[36] As a result of such findings, it has been proposed that the leadership prototype paradigm be updated to Think Manager, Think Heterosexual Male.[37]

One explanation for the development of inverted gender stereotypes is that we have a tendency to exaggerate similarities among people in "out-groups"—that is, people we don't share similarities with and therefore don't typically associate with.[38] For example, as lesbians and heterosexual men have the same sexual attraction to women, it is assumed that lesbians must be more similar to heterosexual men than they are to heterosexual women[39] and therefore become stereotyped as gender-atypical.[40] This binary conceptualisation of sexuality, based in large part on the perceived gender of a person's partner, could also explain why stereotypes of other sexual minorities such as bisexual, pansexual and asexual individuals are less gendered. Another possible explanation for the development of inverted gender assumptions is that we see gender and sexuality as shaping each other, insofar as part of what it means to be a man or masculine is to have heterosexual relationships and desire.[41] The expectation is therefore that lesbians and gay men are gender-atypical—and gender-atypical people must identify as lesbian or gay.

These typical representations of gay men and lesbians, also known as archetypes, have been perpetuated and disseminated through their reproduction in the media. Gay men are often typecast as effeminate and lesbian women are often associated with masculine traits.[42] Historically TV and film have used non-gender-conforming behaviours or interest alone to instil an assumption in the audience that the character is not heterosexual.

So far we have focused on the stereotypes of lesbians and gay men which are only part of the story as there are many other identities within the community that we haven't yet explored. It is important to recognise that there is a relative lack of research and so a lot less is known about the stereotypes we hold about bisexual and transgender individuals. What we do know, however, is that stereotypes for these identities are quite different from those for lesbians and gay men.

The stereotypes held about bisexual individuals are generally negative[43] with individuals commonly seen as "disingenuous," "dishonest," "indecisive," "confused" and "selfish." The stereotypes of bisexual men and women show evidence of sexual double standards, with bisexual women being viewed as more promiscuous than bisexual men. The stereotypes are also shown to be held by heterosexual, lesbian and gay communities.[44]

Stereotypes associated with transgender people reveal a discomfort with acceptance, being viewed as abnormal, having mental illness[45] and gender ambiguity, such as the idea that transgender women are not really women[46] and that transgender men have both feminine and masculine traits—making them, for example, emotionally strong as well as aggressive. The overall effect of the views held of transgender people is one of treating them as outcasts and of creating social distance from them.[47] Archetypes of transgender people are in part formed through media representation where their characters are typically associated with narratives of fear (criminals including murderers), comic relief and an inability to conform to feminine beauty standards.[48]

Studies have also suggested that male-to-female transgender participants reported their skills and abilities becoming devalued after their transition,[49] whereas female-to-male transgender respondents reported an increase in perceived authority and respect.[50] These effects suggest that stereotypes based on gender expression are in action in these situations.

Stigmatised Identities

The classic work on stigma was produced by Erving Goffman in *Stigma: Notes on the Management of Spoiled Identity*. The word "stigma" comes from Greek and refers to the ways in which someone would be marked as having done something which was disapproved of. This mark needed to be visible easily to others, such as a cut or a burn, and which would warn others of the lower status of this person. They would be identified as "a blemished person, richly polluted, to be avoided, especially in public."[51]

Goffman identified three types of stigma: physical, tribal and character. Physical stigma includes visible disabilities or disfigurement. Tribal stigma refers to race, nationality and religious groups. Character stigma includes people being seen as weak-willed, dishonest and controlled by "domineering, or unnatural passions"[52] because they are addicted to drugs or alcohol and are unemployed, homeless or homosexual. Psychiatry bears a major responsibility for its part in creating the myth that LGBTQ+ people are unnatural and have inherent personality and character defects. Goffman mischievously refers to people not in possession of the stigmatised identity as "normals."

With invisible stigma there is constant pressure on the individual to ensure, they do not inadvertently reveal this tainted side of themselves when interacting with others: "to display or not to display; to tell or not to tell; to let on, or not to let on; to lie or not to lie; and in each case, to whom, how, when, and where."[53] A problem with non-disclosure is "in-deeper-ism": that is, "pressure to elaborate a little further and further to prevent a given disclosure."[54]

Researchers have looked at the function that stigmatisation serves.[55]

They conclude that firstly stigmatisation is a way of exerting domination over people. Secondly, it's a way of maintaining social norms. Thirdly, it's a way of avoiding people who may carry a disease harmful to others. They refer to these three functions as "keeping people down," "keeping people in" and "keeping people away."[56]

The Impact of Stigma

Being stigmatised impacts people in a number of ways:

Enacted stigma—the reactions of others to the stigmatised person including discrimination and exclusion. This form of stigma will be covered in the next chapter. Here we will discuss:

- Felt stigma—the emotions associated with being a member of a discredited group.
- Anticipated stigma—how people from a discredited group will be wary of the situations they are in, watching for signs of antagonism towards them.
- Internalised stigma—also known as internalised homophobia.
- Stigma by association—how others who are not from the stigmatised group can be viewed if they befriend people from the group.

Felt Stigma

Felt stigma refers to the emotions associated with being a member of a discredited group. including shame, fear and disgust.[57] People who conceal a stigmatised identity experience greater stress with consequences that include negative self-evaluations, increased anxiety, higher levels of hostility and depression. "Whereas individuals with a visible stigma face the emotional stress of being devalued, individuals with a concealable stigma must choose between this stress and the emotional stress of hiding."[58] These are powerful feelings with significant negative consequences.

Impression management and self-monitoring become important, with people trying to ensure they don't give any clues to the hidden part of themselves. Concealing their sexual orientation means not sharing information about themselves and can also include lying, which causes more distress. Research among gay male students[59] found that they would adapt the way they talked and what they talked about in order to avoid the rejection they feared would follow if they were open about their sexual orientation.

Another consequence of the fear of being identified is that people may avoid certain social situations and be wary of close relationships with others, which will have a detrimental effect on their well-being. Ironically, in order to conceal an identity a person has to be aware of it at all times. It will mean avoiding conversations and interactions which could be uncomfortable at best and humiliating at worst. This means there is always the danger that it will intrude into their thinking, leading them to suppress those thoughts. The cycle of secrecy, suppression, intrusion and further suppression generates a preoccupation with the nature of the stigma, which inevitably impacts people's well-being and self-identification.

Anticipated Stigma

Anticipated stigma describes the situation where people watch for signs that they will be discriminated against. It means people will be constantly on their guard in social situations, keeping their defences up and watching for any signs of antagonism towards them.[60]

Having a stigma that is not readily identifiable by others creates other dilemmas which in turn lead to increased stress. In every new encounter the individual is making a decision about whether to reveal or conceal. In many situations it will be clear that there is no need to reveal: the interaction may be brief, or you may never meet the person again. In situations where someone gets to know the other person the dilemma may gain more potency, and they may devote more consideration to telling the other person about their identity. The longer one leaves it the harder it becomes to disclose, and the reaction of the other person potentially becomes more negative since they may feel they have been cheated or duped in the earlier stages of the encounter or relationship. By way of contrast "normals" don't have to engage in actively monitoring their environment and therefore feel wholly present in the situation.

Various theories have been developed to discuss concealed identities. Sandra Petronio's communication privacy management (CPM) theory recognises the balancing act between privacy and revealing:

> Privacy has importance for us because it lets us feel separate from others. It gives us a sense that we are the rightful owners of information about us. There are risks that include making private disclosures to the wrong people, disclosing at a bad time, telling too much about ourselves, or compromising others.[61]

We will choose to keep some information entirely private while sharing some with a small group of people so that it becomes part of a collective privacy. This entails a high degree of trust with an inner circle, and there are painful consequences should someone unwittingly reveal the private information without the consent of the individual concerned. Sharing is a difficult decision to make because once information is shared there's no going back.

This is one reason why response rates to monitoring surveys within organisations tend to be much lower for questions relating to sexual orientation. Despite the surveys being anonymous, people will not take the risk of replying because they don't trust the process enough. Mistakes can happen: a person's identity was inadvertently revealed to a large group of people in one of our client organisations. Staff had responded to questions about sexual orientation and also one about seniority. Not only was the person in question easily identified through their seniority but the information was accidentally emailed to a wider group. The mistake was picked up quickly and it was hoped that no damage was done.

Perspective taking, or the ability to understand how another person is thinking and feeling about a situation, is generally a great skill to possess, but in circumstances of anticipated stigma, it heightens the sense of threat that a person may be feeling. It also means they are anticipating and preparing for events in a way that a non-stigmatised person would not be doing. This can extend to increased questioning of the other person's motives. People who have revealed their identity won't be overly concerned about this, but studies have shown that those who have concealed themselves are more suspicious about the motivation of those they are interacting with. They are likely to have a more negative assessment of others and a higher degree of paranoia about the assessments they believe others are making of them.[62]

The impact on people being stigmatised at work is considerable: they are more likely to state they wish to leave the organisation,[63] more likely to be discriminated against if they have come out than if they have not,[64] have less confidence that they will be treated fairly in the workplace and have greater expectation that decisions about them will be made on characteristics unrelated to work performance. Compared to their straight colleagues, they also feel they are less likely to be promoted and to achieve their full potential in their roles.[65]

Internalised Stigma

Internalised stigma is also known as internalised homophobia. The term "homophobia" was introduced in 1972 by psychologist George Weinberg in his book *Society and the Healthy Homosexual* (Weinberg, G.H., 1972 Macmillan)[66] and then actively campaigned to get the word accepted.[67] It incorporated the prejudice people faced as well as the self-loathing that they experience. Because of the discrimination and exclusion they experience, people can start to accept the negative views others have of them. This has a detrimental impact on well-being, self-esteem and self-efficacy. It can also lead to depression and an increased likelihood of suicide. Knowing society's negative evaluation of a stigma, having to be careful about revealing even small details about yourself, constantly monitoring situations for signs that your identity may be exposed—all these burdens have a negative impact on people's self-perception.

The sense of being stigmatised starts from an early age and LGBTQ+ youth experience homophobia in their relationships with family, peers and caregivers. However, this experience is modulated by a number of factors including location, ethnicity and religion. If someone is living in an area where religion plays a great part in people's daily lives, then they are more likely to experience what is known as structural stigma, where societal norms are embedded into anti-LGBTQ+ policies.[68]

Those with a visible stigmatised identity may feel that being able to conceal such an identity makes for an easier life. The avoidance of overt hostility, harassment and discrimination must surely make navigating your way through interactions much more comfortable. However, the research shows clearly that the constant battle not to reveal oneself, excessive rumination on the stigmatised identity and avoidance of social interactions where more support could be available all have a huge impact on people's view of themselves. Studies examining other non-visible identities such as class can shed light on how LGBTQ+ individuals will be feeling. Working-class students at elite universities can present themselves to fellow students as coming from more privileged backgrounds so as to enable productive interactions. On the surface this would appear to be a successful strategy—but it leads to feelings of guilt, inauthenticity and fraudulence. Identity ambivalence[69] also occurs, where they feel they have betrayed their friends and families by choosing to deny their roots. Not sharing one's identity with others also deprives people of the company of others with the same characteristic which leads to a loss of self-esteem.

These factors combine to make individuals feel that they are less able to control events around them and carry out tasks effectively—otherwise known as self-efficacy. The combination of these behaviours, thoughts and feelings can become part of a vicious cycle:

> a negative self-image can impact cognitions (e.g., "my secret stigma is shameful and I am a worse person for possessing it"), affect (e.g., depression, anxiety), and behaviour (e.g., impression management attempts to conceal the shameful stigma, the avoidance of situations likely to heighten shame-related cognition and affect).[70]

In a major review of studies looking at sexual concealment and its impact on mental health, the authors concluded that there is a relationship between the two. The minority stress model is applicable here. This model shows that people who are in a minority experience pressure which more dominant groups do not. The relationship between concealment and mental health problems "was larger for those studies that assessed concealment as lack of open behaviour, those conducted recently, and those with younger samples; it was smaller in exclusively bisexual samples."[71]

Stigma by Association

In a meeting between three PRIDE employee resource group members and the company's Chief Operating Officer, they asked if he would act as champion for the ERG. Clearly startled, he asked: "But why me?" As the conversation continued his tension level noticeably reduced. The situation was part of an inclusive leadership programme we were running, and the PRIDE network members were in fact actors. But the COO was playing himself and his reactions were genuine. During the debrief he was open and candid enough to recognise that his first reaction on being asked to take a high-profile role for the PRIDE group was one of shock. He had experienced another stigma-related phenomenon: stigma by association.

Neuberg and colleagues make the interesting observation that the cost of stigmatisation isn't just felt by the people who are the victims of it but also by the people who do it: "to stigmatise another is to paint that person with broad strokes, often obscuring positive qualities that may be of great value."[72] In carefully designed experiments they sought to discover what happens when stigmatised people are with "normal" people. In one experiment participants were told that they were going to view a conversation between two friends which they would be observing via closed circuit television (in fact it was a recording). They were also provided with information about the two people engaged in the interaction including their sexual orientation and occupational status. Having watched the interaction, they would then rate the two people involved.

Participants judged the person they were observing more harshly if they thought they were gay, seeing them as more dishonest and untrustworthy.

The researchers also found "a strong stigmatisation by association effect, heterosexual targets being derogated as a consequence of their association with the homosexual targets".[73] In other words a heterosexual person interacting with a homosexual person was viewed more negatively than if the two people were heterosexual. They also found that if the person was of high status the level of stigmatisation increased.

The study highlighted the deep emotional reaction that some people have towards people with a minority sexual orientation. This reaction is so strong that it can affect the evaluations made of people who are heterosexual but who engage in friendships and interactions with people who are gay. The wariness felt by our senior leader when asked to be the champion for the PRIDE network therefore has some justification in research. He may indeed have been viewed more negatively by being involved with the network than if he remained separate from it.[74] When LGBTQ+ rights are being publicly discussed, stress increases for people from a minority sexual orientation. However, stress also increases for heterosexual allies who feel the impact of such debates in the same way that the minorities do.[75,76]

Related to stigma by association is courtesy stigma. This refers to non-stigmatised individuals (or "normals") treating the stigmatised person as if they didn't have the stigma. This can be a positive behaviour but it is liable to stresses and strains which undermine its intent:

> the person with a courtesy stigma can in fact make both the stigmatised and the normal uncomfortable: by always being ready to carry a burden that is not "really" theirs, they can confront everyone else with too much morality; by treating the stigma as a neutral matter to be looked at in a direct, off-hand way, they open themselves and the stigmatised to misunderstanding by normals who may read offensiveness into this behaviour.[77]

It is like saying to the person: "You are just like one of us" which could also mean that: "You are not like the rest of them". The person treated in this way may doubt that they have really been accepted, even on a courtesy basis. This will particularly apply if the others who have extended the courtesy do not truly understand or accept the level of discrimination which the person faces. In doing so, they can trivialise the life experiences of the stigmatised individual.

How to Cope with LGBTQ+ Stigma

Researchers Richard Lazarus and Susan Folkman's classic work on stress identified two broad coping strategies which they called problem-focused (analytical and rational) and emotion-focused (dealing with the feelings related to the stress being experienced).[78] It was important that people identified the right approach for their situation or things would not improve for

Table 4.1 Interaction between levels of disclosure and empowerment

	Low empowerment	*High empowerment*
Full disclosure	Telling some people in the organisation e.g. HR Dealing with issues as they arise e.g. harassment Sharing after an important moment e.g. passing exams—when people feel more secure	Sharing with colleagues Sharing more widely Coming more involved in LGBTQ+ inclusion Being an advocate Being seen as a role model
No disclosure	Dealing with situations alone Monitoring of others' behaviour Seeking to be a star performer to avoid criticism Alcohol and drug abuse	Seek support outside of organisation, including health professionals, legal professionals and their religion

them. It has proven to be an influential approach. However, it's not as clear cut as it sounds especially where someone's identity is not apparent to others. An LGBTQ+ employee may be more accepted as part of a team if they conceal their real identity, but this could be at the cost of feeling less included and having a reduced sense of belonging to the organisation.

Another coping model for LGBTQ+ employees looks at the interaction between two factors[79] One is the level of disclosure (from no disclosure to full disclosure) and the other is the level of empowerment (from low—having to deal with discrimination on one's own—to high, where there is support from the organisation and colleagues). As can be seen in Table 4.1, this leads to four broad approaches.[80]

Internal Strategies (Low Concealment, Low Empowerment)

Where there is little or no disclosure and low levels of empowerment, individuals will try to deal with situations on their own. The fact that people have not disclosed will mean that they are not able to access support from other LGBTQ+ colleagues either. The approaches include:

- Preventative-preparative strategies, which include avoiding situations where their identity might be revealed, monitoring other people's behaviours and self-regulating their behaviour and topics of conversation. These strategies also include choice of profession, sector and employer with preference being given to those which are more inclusive.
- Outperforming in their job, which includes being outstanding in the work they do, taking on difficult challenges and being seen as a star performer.

- Disengagement—not being fully involved in their work and looking to move to a different area or leave the organisation.
- Harmful coping strategies such as drinking too much and drug abuse.

External Strategies (Low Concealment, High Empowerment)

Strategies in this quadrant include:

- Turning for support from people outside of the organisation.
- Seeking advice from health professionals and the legal profession.
- Spiritual and religious direction, which may come from a religious community or from their relationship with their god or gods. In particularly religious countries where anti-LGBTQ+ attitudes are strong, it is sometimes the religion itself which provides a source of comfort for individuals by focusing on the narratives of love that are contained within the traditions of the religion: for example

  they use the same Bible to crush the LGBTQ+ family, telling them they'll go to hell, they'll burn, that's why Gomorrah and Sodom was destroyed because of such habits, why do this, why don't you do that and I think it's wrong because nowhere in the Bible does it say you should love this one and hate this one. It's all about love. I think the love part should be read other than the hate part.[81]

Reactive Strategies (High Concealment, Low Empowerment)

These strategies include:

- Informing some people such as HR but not letting their identity be known more publicly.
- Coming out only once they feel secure in their position; for example, after they have completed professional exams or have reached a level of seniority where they feel safer.
- Dealing with individual situations when they arise, such as confronting a harasser directly.

Proactive Strategies (High Concealment, High Empowerment)

Proactive strategies include:

- Disclosing their identity to close colleagues.
- Making their identity known more widely.
- Being involved in building an LGBTQ+ inclusive environment by ensuring fairness of treatment.

- Acting as an advocate for LGBTQ+ staff.
- Being seen as a role model for others.

In order for proactive strategies to be possible the organisation needs to have created an environment where people feel valued, trusted and included. It also needs to be a workplace where, if someone were to advocate for LGBTQ+ staff, this would be viewed positively and not as a source of criticism.

Well-being Strategies

The creators of this model[82] refer to well-being strategies under low conceal-ment and low empowerment. However, we feel that this is a cross-category set of strategies which anybody in any category can employ. These strategies include:

- Emotional-regulation strategies—the focus here is on how to handle dis-crimination when it is experienced and to build resilience and long-term well-being through mindfulness techniques and exercises.
- Self-affirmation—this involves exercises which increase self-esteem and self-confidence, identifying strengths and seeking to build on them.
- Cognitive reframing—where people try to understand the discrimination, they have experienced by looking at it from different angles, including try-ing to understand the perpetrators.

Strategies like these are also known as stress-related growth. Adopting these practices has many beneficial outcomes such as a more positive self-image and better sense of connection with people who share the same identity.[83]

Social psychologist Colette van Laar and her colleagues identify the organ-isational climate and culture as particularly important in helping individu-als counter the effects of being stigmatised at work.[84] The researchers bring together two concepts: identity threat and psychological safety to create together identity safety. Where the organisation affirms different identities and has policies which support and value all employees, identity threat is less of an issue. However, the organisation must also make it clear that it values white men in the organisation just as much in order to minimise the backlash that can occur if the focus is continually on marginalised groups. Where people feel that support is available to them in the organisation, not only will they feel greater engagement and a higher sense of belonging but organisational results will also be better.

Key Points

Stereotypes are the oversimplified, preconceived notions or beliefs that we hold about certain groups of people. They are often based on generalisations of limited information.

It's important when we consider stereotypes of LGBTQ+ that we consider each identity separately. While there is overlap there are also differences between each of these areas.

The widely held stereotypes of men and women are typically reversed when considering gay men and lesbians. Gay men are seen as stereotypically more effeminate with lesbians being associated with more masculine stereotypes. People who are bisexual are stereotypically seen as dishonest and confused people who are not to be trusted. Transgender individuals are seen and treated almost as if they are aliens.

The impact of othering individuals like this is to stigmatise them—to mark them as different and inferior from those that Ervin Goffman referred to as the "normals." Four types of stigma were covered in this chapter (enacted stigma is covered in Chapter 5):

- Felt stigma—the emotions associated with being a member of a discredited group.
- Anticipated stigma—how people from a discredited group will be wary of the situations they are in, watching for signs of antagonism towards them.
- Internalised stigma—also known as internalised homophobia.
- Stigma by association—how others who are not from the stigmatised group can be viewed if they befriend people from the group.

One effect of being stigmatised with an identity that is not visible to others is that people will choose to conceal it. This has consequences in terms of people's self-image, self-esteem and self-efficacy. It has consequences in terms of people's psychological well-being.

Strategies were shared on how to limit the damaging impact of being stigmatised. The ideal situation in an organisation is born in which people feel empowered to take action themselves which impacts the level of disclosure.

In addition, the "normals" can take action to manage their own biases which includes being able to take the perspective of those being judged.

Notes

1 Epitropaki, O., & Martin, R. (2004). Implicit Leadership Theories in Applied Settings: Factor Structure, Generalizability, and Stability Over Time. *Journal of Applied Psychology*, 89(2), pp. 293–310. https://doi.org/10.1037/0021-9010.89.2.293
2 Burgess, D., & Borgida, E. (1999). Who Women are, Who Women Should be: Descriptive and Prescriptive Gender Stereotyping in Sex Discrimination. *Psychology, Public Policy, and Law*, 5, pp. 665–692.
3 Eagly, A.H., & Karau, S.J. (2002). Role Congruity Theory of Prejudice Toward Female Leaders. *Psychological Review*, 109(3), pp. 573–598. https://doi.org/10.1037//0033-295X.109.3.573
4 Heilman, M.E. (2001). Description and Prescription: How Gender Stereotypes Prevent Women's Ascent Up the Organizational Ladder. *Journal of Social Issues*, 57, pp. 657–674.

5 Eagly, A.H. (1987). *Sex Differences in Social Behavior: A Social-Role Interpretation*. Hillsdale, NJ: Erlbaum.

6 Sczesny, S., Spreemann, S., & Stahlberg, D. (2006). Masculine = Competent? Physical Appearance and Sex as Sources of Gender-Stereotypic Attributions. *Swiss Journal of Psychology*, 65(1), pp. 15–23. https://doi.org/10.1024/1421–0185.65.1.15

7 Schein, V.E., Mueller, R., Lituchy, T., & Liu, J. (1996). Think Manager-Think Male: A Global Phenomenon? *Journal of Organizational Behavior*, 17, pp. 33–41.

8 Eagly, A.H., & Carli, L.L. (2007). *Through the Labyrinth: The Truth about How Women become Leaders*. Boston, MA: Harvard Business School Press.

9 Braun, S., Stegmann, S., Hernandez Bark, A.S., Junker, N.M., & van Dick, R. (2017). Think Manager—Think Male, Think Follower—Think Female: Gender Bias in Implicit Followership Theories. *Journal of Applied Social Psychology*, 47(7), pp. 377–388.

10 Barrantes, R.J., & Eaton, A.A. (2018). Sexual Orientation and Leadership Suitability: How Being a Gay Man Affects Perceptions of Fit in Gender-Stereotyped Positions. *Sex Roles*, pp. 1–16.

11 Rudman, L.A. (1998). Self-Promotion as a Risk Factor for Women: The Costs and Benefits of Counterstereotypical Impression Management. *Journal of Personality and Social Psychology*, 74, pp. 629–645.

12 Heilman, M.E., Wallen, A.A., Fuchs, D., & Tamkins, M.M. (2004). Penalties for Success: Reactions to Women Who Succeed at Male Gender-Typed Tasks. *Journal of Applied Psychology*, 89, pp. 416–427.

13 Steele, C.M., & Aronson, J. (1995). Stereotype Threat and the Intellectual Test Performance of African Americans. *Journal of Personality and Social Psychology*, 69(5), pp. 797–811. https://doi.org/10.1037/0022-3514.69.5.797

14 Davies, P.G., Spencer, S.J., & Steele, C.M. (2005). Clearing the Air: Identity Safety Moderates the Effects of Stereotype Threat on Women's Leadership Aspirations. *Journal of Personality and Social Psychology*, 88(2), pp. 276–287. https://doi.org/10.1037/0022-3514.88.2.276

15 Koenig, A.M., Eagly, A.H., Mitchell, A., & Ristikari, T. (2011). Are Leader Stereotypes Masculine? A Meta-Analysis of Three Research Paradigms. *Psychological Bulletin*, 137(4), pp. 616–642. https://doi.org/10.1037/a0023557

16 Cochran, A., Hauschild, T., Elder, W.B., Neumayer, L.A., Brasel, K.J., & Crandall, M.L. (2013). Perceived Gender-based Barriers to Careers in Academic Surgery. *American Journal of Surgery*, 206(2), pp. 263–268. https://doi.org/10.1016/j.amjsurg.2012.07.044

17 Blickenstaff, J.C. (2005). Women and Science Careers: Leaky Pipeline or Gender Filter? *Gender and Education*, 17(4), pp. 369–386.

18 Sellers, N., Satcher, J., & Comas, R. (1999). Children's Occupational Aspirations: Comparisons by Gender, Gender Role Identity, and Socioeconomic Status. *Professional School Counseling*, 2, pp. 314–317.

19 Levitt, E.E., & Klassen, A.D. (1974). Public Attitudes Toward Homosexuality: Part of the 1970 National Survey by the Institute for Sex Research. *Journal of Homosexuality*, 1, pp. 29–43.

20 Madon, S. (1997). What Do People Believe about Gay Males? A Study of Stereotype Content and Strength. *Sex Roles*, 37, pp. 663–685.

21 Herek, G.M. (1988). Heterosexuals' Attitudes Toward Lesbians and Gay Men: Correlates and Gender Differences. *Journal of Sex Research*, 25, pp. 451–477.

22 Kite, M. (1992). Psychometric Properties of the Homosexuality Attitude Scale. *Representative Research in Social Psychology*, 19, pp. 79–94.

23 Morrison, M.A., & Morrison, T.G. (2002). Development and Validation of a Scale Measuring Modern Prejudice Toward Gay Men and Lesbian Women. *Journal of Homosexuality*, 43(2), pp. 15–37.

24 Nosek, B.A., Smyth, F.L., Hansen, J.J., Devos, T., Lindner, N.M., Ranganath, K.A., . . . Banaji, M.R. (2007). Pervasiveness and Correlates of Implicit Attitudes and Stereotypes. *European Review of Social Psychology*, 18, pp. 36–88.

25 Sabin, J.A., Riskind, R.G., & Nosek, B.A. (2015). Health Care Providers' Implicit and Explicit Attitudes Toward Lesbian Women and Gay Men. *American Journal of Public Health*, 105(9), pp. 1831–1841. https://doi.org/10.2105/AJPH.2015.302631

26 Blashill, A.J., & Powlishta, K.K. (2009). Gay Stereotypes: The Use of Sexual Orientation as a Cue for Gender-Related Attributes. *Sex Roles*, 61, pp. 783–793.

27 Kite, M.E., & Deaux, K. (1987). Gender Belief Systems: Homosexuality and the Implicit Inversion Theory. *Psychology of Women Quarterly*, 11, pp. 83–96.

28 Clarke, H.M., & Arnold, K.A. (2017). Diversity in Gender Stereotypes? A Comparison of Heterosexual, Gay and Lesbian Perspectives. *Canadian Journal of Administrative Sciences*, 34, pp. 149–158. https://doi.org/10.1002/cjas.1437

29 Vaughn, A.A., Teeters, S.A., Sadler, M.S., & Cronan, S.B. (2017). Stereotypes, Emotions, and Behaviors Toward Lesbians, Gay Men, Bisexual Women, and Bisexual Men. *Journal of Homosexuality*, 64(13), pp. 1890–1911. https://doi.org/10.1080/00918369.2016.1273718

30 Barrantes, R.J., & Eaton, A.A. (2018). Sexual Orientation and Leadership Suitability: How Being a Gay Man Affects Perceptions of Fit in Gender-Stereotyped Positions. *Sex Roles*, pp. 1–16.

31 Clarke, H.M., & Arnold, K.A. (2017). Diversity in Gender Stereotypes? A Comparison of Heterosexual, Gay and Lesbian Perspectives. *Canadian Journal of Administrative Sciences*, 34, pp. 149–158. https://doi.org/10.1002/cjas.1437

32 Brambilla, M., Carnaghi, A., & Ravenna, M. (2011). Status and Cooperation Shape Lesbian Stereotypes: Testing Predictions from the Stereotype Content Mode. *Social Psychology*, 42(2), pp. 101–110.

33 Clausell, E., & Fiske, S.T. (2005). When Do Subgroup Parts Add Up to the Stereotypic Whole? Mixed Stereotype Content for Gay Male Subgroups Explains Overall Ratings. *Social Cognition*, 23(2), pp. 161–181.

34 Geiger, W., Harwood, J., & Hummert, M.L. (2006). College Students' Multiple Stereotypes of Lesbians: A Cognitive Perspective. *Journal of Homosexuality*, 51(3), pp. 165–182.

35 Duggan, L. (2002). The New Homonormativity: The Sexual Politics of Neoliberalism. In: R. Castronovo, D. Nelson, & D. Pease (eds.), *Materializing Democracy: Toward a Revitalized Cultural Politics*. New York, USA: Duke University Press, pp. 175–194. https://doi.org/10.1515/9780822383901-008

36 Liberman, B.E., & Golom, F.D. (2015). Think Manager, Think Male? Heterosexuals' Stereotypes of Gay and Lesbian Managers. *Equality, Diversity and Inclusion: An International Journal*, 34(7), pp. 566–578.

37 Morton, J.W. (2017). Think Leader, Think Heterosexual Male? The Perceived Leadership Effectiveness of Gay Male Leaders. *Canadian Journal of Administrative Sciences*, 34(2), pp. 159–169. https://doi.org/10.1002/cjas.1434

38 Wilder, D.A. (1981). Perceiving Persons as a Group: Categorization and Intergroup Relations. In: *Cognitive Processes in Stereotyping and Intergroup Behavior*. Hillsdale, NJ: Erlbaum, pp. 213–257.

39 Barrantes, R.J., & Eaton, A.A. (2018). Sexual Orientation and Leadership Suitability: How Being a Gay Man Affects Perceptions of Fit in Gender-Stereotyped Positions. *Sex Roles*, pp. 1–16.

40 Blashill, A.J., & Powlishta, K.K. (2009). Gay Stereotypes: The Use of Sexual Orientation as a Cue for Gender-Related Attributes. *Sex Roles*, 61, pp. 783–793.

41 Eaton, A.A., & Matamala, A. (2014). The Relationship between Heteronormative Beliefs and Verbal Sexual Coercion in College Students. *Archives of Sexual Behavior*, 43(7), pp. 1443–1457.

42 Raley, A.B., & Lucas, J.L. (2006). Stereotype or Success? Prime-Time Television's Portrayals of Gay Male, Lesbian, and Bisexual Characters. *Journal of Homosexuality*, 51(2), pp. 19–38.

43 Mize, T.D., & Manago, B. (2018). The Stereotype Content of Sexual Orientation. *Social Currents*, 5(5), pp. 458–478. https://doi.org/10.1177/2329496518761999

44 Monro, S. (2015). *Bisexuality: Identities, Politics, and Theories*. Basingstoke, England: Palgrave MacMillan.

45 Howansky, K., Wilton, L., Young, D., Abrams, S., & Clapham, R. (2019). (Trans) Gender Stereotypes and the Self: Content and Consequences of Gender Identity Stereotypes. *Self and Identity*, 20, pp. 1–18. 10.1080/15298868.2019.1617191.

46 Billard, T.J. (2018). Attitudes Toward Transgender Men and Women: Development and Validation of a New Measure. *Frontiers in Psychology*, 9, Article 387. https://doi.org/10.3389/fpsyg.2018.00387

47 Gazzola, S.B., & Morrison, M.A. (2014). Cultural and Personally Endorsed Stereotypes of Transgender Men and Transgender Women: Notable Correspondence or Disjunction? *International Journal of Transgenderism*, 15(2), pp. 76–99.

48 Wellborn, C. (2015). *Reactions of the Transgender Community Regarding Media Representation*. Arlington, TX: University of Texas.

49 Schilt, K., & Connell, C. (2007). Do Workplace Gender Transitions Make Gender Trouble? *Gender, Work and Organization*, 14(6), pp. 596–618. https://doi.org/10.1111/j.1468-0432.2007.00373.x

50 Griggs, C. (1998). *S/he: Changing Sex and Changing Clothes*. New York: Berg.

51 Goffman, E. (1963). *Stigma: Notes on the Management of Spoiled Identity*. Englewood Cliffs, NJ: Prentice-Hall, p. 16.

52 Goffman, E. (1963). *Stigma: Notes on the Management of Spoiled Identity*. Englewood Cliffs, NJ: Prentice-Hall, p. 4.

53 Goffman, E. (1963). *Stigma: Notes on the Management of Spoiled Identity*. Englewood Cliffs, NJ: Prentice-Hall, p. 42.

54 Goffman, E. (1963). *Stigma: Notes on the Management of Spoiled Identity*. Englewood Cliffs, NJ: Prentice-Hall, p. 83.

55 Phelan, J.C., Link, B.G., & Dovidio, J.F. (2008). Stigma and Discrimination: One Animal or Two? *Social Science & Medicine*, 67, pp. 358–367.

56 Phelan, J.C., Link, B.G., & Dovidio, J.F. (2008). Stigma and Discrimination: One Animal or Two? *Social Science & Medicine*, 67, pp. 358–367.

57 Parker, C.M., Hirsch, J.S., Philbin, M.M., & Parker, R.G. (2018, October). The Urgent Need for Research and Interventions to Address Family-Based Stigma and Discrimination Against Lesbian, Gay, Bisexual, Transgender, and Queer Youth. *Journal of Adolescent Health*, 63(4), pp. 383–393. https://doi.org/10.1016/j.jadohealth.2018.05.018. Epub 2018 August 23. PMID: 30146436; PMCID: PMC6344929

58 Pachankis, J.E. (2007, March). The Psychological Implications of Concealing a Stigma: A Cognitive-Affective-Behavioral Model. *Psychological Bulletin*, 133(2), pp. 328–345. https://doi.org/10.1037/0033-2909.133.2.328. PMID: 17338603

59 Pachankis, J.E., & Goldfried, M.R. (2006). Social Anxiety in Young Gay Men. *Journal of Anxiety Disorders*, 20(8), pp. 996–1015. https://doi.org/10.1016/j.janxdis.2006.01.001

60 Parker, C.M., Hirsch, J.S., Philbin, M.M., Parker, R.G. (2018, October). The Urgent Need for Research and Interventions to Address Family-Based Stigma and Discrimination Against Lesbian, Gay, Bisexual, Transgender, and Queer Youth. *Journal of Adolescent Health*, 63(4), pp. 383–393. https://doi.org/10.1016/j.jadohealth.2018.05.018. Epub 2018 August 23. PMID: 30146436; PMCID: PMC6344929.

61 Petronio, S. (2002). *Boundaries of Privacy: Dialectics of Disclosure*. State University of New York Press, ProQuest Ebook Central.

62 Pachankis, J.E. (2007). The Psychological Implications of Concealing a Stigma: A Cognitive–Affective–Behavioral Model. *Psychological Bulletin*, 133(2), pp. 328–345.

63 Ragins, B.R., & Cornwell, J.M. (2001). Pink Triangles: Antecedents and Consequences of Perceived Workplace Discrimination Against Gay and Lesbian Employees. *Journal of Applied Psychology*, 86, pp. 1244–1261.

64 Sears, B., & Mallory, C. (2011). *Documented Evidence of Employ-ment Discrimination & Its Effects on LGBT People*. The Williams Institute. https://williamsinstitute.law.ucla.edu/wp-content/uploads/Sears-Mallory-Discrimination-July-2 0111.pdf

65 Lewis, G.B., & Pitts, D.W. (2017). LGBT–Heterosexual Differences in Perceptions of Fair Treatment in the Federal Service. *The American Review of Public Administration*, 47, pp. 574–587.

66 Weinberg, G.H. (1972). *Society and the Healthy Homosexual*. Macmillan.

67 Drescher, J. (2016). An Interview with George Weinberg, PhD. *Journal of Gay & Lesbian Mental Health*, 20(1), pp. 87–93.

68 Parker, C.M., Hirsch, J.S., Philbin, M.M., & Parker, R.G. (2018, October). The Urgent Need for Research and Interventions to Address Family-Based Stigma and Discrimination Against Lesbian, Gay, Bisexual, Transgender, and Queer Youth. *Journal of Adolescent Health*, 63(4), pp. 383–393. https://doi.org/10.1016/j.jadohealth.2018.05.018. Epub 2018 August 23. PMID: 30146436; PMCID: PMC6344929

69 Pachankis, J.E. (2007). The Psychological Implications of Concealing a Stigma: A Cognitive–Affective–Behavioral Model. *Psychological Bulletin*, 133(2), pp. 328–345.

70 Pachankis, J.E. (2007). The Psychological Implications of Concealing a Stigma: A Cognitive–Affective–Behavioral Model. *Psychological Bulletin*, 133(2), pp. 328–345.

71 Pachankis, J.E., Mahon, C.P., Jackson, S.D., Fetzner, B.K., & Bränström, R. (2020). Sexual Orientation Concealment and Mental Health: A Conceptual and Meta-Analytic Review. *Psychological Bulletin*, 146(10), p. 831.

72 Neuberg, S.L., Smith, D.M., Hoffman, J.C., & Russell, F.J. (1994). When We Observe Stigmatized and "Normal" Individuals Interacting: Stigma by Association. *Personality & Social Psychology Bulletin*, 20(2), pp. 196–209.

73 Neuberg, S.L., Smith, D.M., Hoffman, J.C., & Russell, F.J. (1994). When We Observe Stigmatized and "Normal" Individuals Interacting: Stigma by Association. *Personality & Social Psychology Bulletin*, 20(2), pp. 196–209.

74 Neuberg, S.L., Smith, D.M., Hoffman, J.C., & Russell, F.J. (1994). When We Observe Stigmatized and "Normal" Individuals Interacting: Stigma by Association. *Personality & Social Psychology Bulletin*, 20(2), pp. 196–209.

75 Ecker, S., Riggle, E.D., Rostosky, S.S., & Byrnes, J.M. (2019). Impact of the Australian Marriage Equality Postal Survey and Debate on Psychological Distress Among Lesbian, Gay, Bisexual, Transgender, Intersex and Queer/Questioning People and Allies. *Australian Journal of Psychology*, 71(3), pp. 285–295. https://doi.org/10.1111/ajpy.12245

76 Anderson, J.R., Campbell, M., & Koc, Y. (2020). A Qualitative Exploration of the Impact of the Marriage Equality Debate on Same-Sex Attracted Australians and Their Allies. *Australian Psychologist*, 55(6), pp. 700–714.

77 Goffman, E. (1963). *Stigma: Notes on the Management of Spoiled Identity*. Englewood Cliffs, NJ: Prentice-Hall, p. 31.

78 Lazarus, R.S., & Folkman, S. (1987). Transactional Theory and Research on Emotions and Coping. *European Journal of Personality*, 1(3), pp. 141–169.

79 Mara, L.-C., Ginieis, M., & Brunet-Icart, I. (2021). Strategies for Coping with LGBT Discrimination at Work: A Systematic Literature Review. *Sexuality Research & Social Policy*, 18(2), pp. 339–354.

80 Mara, L.-C., Ginieis, M., & Brunet-Icart, I. (2021). Strategies for Coping with LGBT Discrimination at Work: A Systematic Literature Review. *Sexuality Research & Social Policy*, 18(2), pp. 339–354.

81 Logie, C.H., Perez-Brumer, A., Woolley, E., Madau, V., Nhlengethwa, W., Newman, P.A., & Baral, S.D. (2018). Exploring Experiences of Heterosexism and Coping Strategies Among Lesbian, Gay, Bisexual, and Transgender Persons in Swaziland. *Gender & Development*, 26(1), pp. 15–32. https://doi.org/10.1080/13552074.2018.1429088

82 Mara, L.-C., Ginieis, M., & Brunet-Icart, I. (2021). Strategies for Coping with LGBT Discrimination at Work: A Systematic Literature Review. *Sexuality Research & Social Policy*, 18(2), pp. 339–354.

83 Wang, K., Rendina, H.J., & Pachankis, J.E. (2016). Looking on the Bright Side of Stigma: How Stress-Related Growth Facilitates Adaptive Coping Among Gay and Bisexual Men. *Journal of Gay & Lesbian Mental Health*, 20(4), pp. 363–375.

84 Van Laar, C., Meeussen, L., Veldman, J., Van Grootel, S., Sterk, N., & Jacobs, C. (2019). Coping with Stigma in the Workplace: Understanding the Role of Threat Regulation, Supportive Factors, and Potential Hidden Costs. *Frontiers in Psychology*, 10, p. 422443.

Chapter 5

Discrimination and Exclusion at Work

Stigma and prejudice towards the LGBTQ+ community have developed from societal attitudes, norms and legislation. But what impact do they have on individuals in terms of their careers? In this chapter, we explore discrimination and exclusion, or what Goffman referred to as enacted stigma which is the reaction of other people towards the stigmatised person and "can manifest as aversion to interaction, avoidance, social rejection, discounting, discrediting, dehumanisation, and depersonalisation of others into stereotypic caricatures."[1]

We unpack how the evolving societal attitudes towards LGBTQ+ individuals impact the employee life cycle: from deciding at a young age which careers to go into to the experiences of discrimination and exclusion of LGBTQ+ employees at work.

Career Planning

The Bottleneck Hypothesis

Sexual and gender identity development typically follow a similar timeline to decisions made about which career to pursue with both occurring in late adolescence and early adulthood.

The bottleneck hypothesis was proposed in 1991 by Hetherington to acknowledge the additional stress that LGB individuals undergo while simultaneously navigating their sexual identity development and attempting to plan a career path.[2] The challenges young people face during this period, particularly in the face of societal stigma and prejudice, create a conflict between the psychological resources needed to navigate this phase and those required for career development.[3]

For example, LGB individuals are more likely to be subjected to bullying which in turn leads to truancy, poorer academic performance and reduced career aspirations.[4] Other areas of impact include increased feelings of shame,[5] lower self-esteem, reduced perceived social support, as well as depression and loneliness.[6] LGBTQ+ adolescents also experience heightened

DOI: 10.4324/9781003489580-7

vigilance of others to ensure physical safety. This comes at the expense of time and energy, detracting from opportunities to study, pursue academic interests, network and engage in work experience.

More recently the bottleneck hypothesis has been used to understand the experiences of transgender individuals.[7] In addition to the concerns that LGB individuals face during this period of identity development, transgender students also face worries about gender presentation, and experience increased levels of distress and victimisation.[8] The psychological stress of this distracts some transgender students from fully engaging with vocational planning and decision making.[9]

Occupational Segregation

One consequence of discrimination is occupational segregation: "the systematic distribution of people across occupations based on demographic characteristics."[10] One of the first people to identify this phenomenon was Havelock Ellis, who noticed that gay men were overrepresented in the arts sector, especially in acting and literature. He also noted other occupations including hairdressing: "I have been told that among London hairdressers' homosexuality is so prevalent that there is even a special attitude which the client may adopt in the chair to make known that he is an invert."[11]

More recent research has found that gay men are more likely to be found in female-dominated occupations with the reverse being true for lesbians. However, there were also other occupations where both gay men and lesbians would be found and "these occupations seem to have little in common, ranging from some blue-collar trades (such as various repairers and mechanics) to service jobs (such as flight attendant and massage therapists) and white-collar occupations (such as psychologists and post-secondary teachers)."[12]

This overrepresentation of gay men and lesbians in gender-atypical work has been reported across various research studies[13] which consistently show that gay men are more likely to be in female-majority occupations than heterosexual men and that lesbians are more likely to hold positions in male-majority occupations than heterosexual women.[14] However, there has been a lack of explanation for such segregation and the suggestions put forward have been speculative and limited to a small set of specific occupations.[15] This means that a large proportion of occupational segregation remains unexplained.

One of the most common reasons given as to why some occupations are more attractive to the LGBTQ+ community is that there are more LGBTQ+ people working in them. The circularity of the argument means that this is more of a description than an explanation. The key here is concealment which is a significant component of stigma management and so people will avoid those occupations where they feel the consequences of revealing their sexual orientation will be greater. In those occupations where greater numbers of

LGBTQ+ staff have disclosed the need for concealment will be less, and the consequences of revealing are more likely to be positive as people will receive support and affirmation.

Another suggestion that sheds light on the career experiences of LGBTQ+ individuals is the safe haven hypothesis.[16] This theory suggests that in identifying a suitable occupation or workplace LGB individuals will seek careers in which they are more likely to be shielded from workplace discrimination. This theory also highlights that once individuals find such a safe haven, they are less likely to leave it. Many researchers have supported the idea that considerations of workplace discrimination play a role in the career decisions of LGBTQ+ individuals.[17] In fact a significant portion of LGBTQ+ career research has centred on experiences or perceptions of discrimination and the need for a secure work environment.[18]

Other researchers have expanded on this theory to pinpoint specific work environments considered safe havens for LGBTQ+ individuals. These settings include the public sector,[19] workplaces with LGBTQ+ friendly policies,[20] as well as self-employment.[21] Favoured occupations include those which align with their social values, have task independence[22] and convey altruism.[23]

Analysis of US datasets from 2008 to 2010 revealed that there were two factors that can explain job segregation:

- Task independence—this relates to a preference for roles that allow individuals to work and perform tasks without substantially depending on co-workers. Working in isolation or at a distance from co-workers may be seen as a strategy to reduce the potential for interpersonal exclusion and discrimination. In these roles the risk of revealing their identity will be reduced and the impact of being excluded is felt less severely. These roles include psychologists, university academics and hairdressers.[24]
- Social perceptiveness—this relates to the accurate anticipation and reading of others' reactions. Monitoring one's own behaviour as well as the behaviour of others in order to avoid being stigmatised means that members of the LGBTQ+ community have a more highly developed sense of social awareness: "awareness and anticipation of others' reactions, and mental states—whether they are patients in healthcare settings, passengers on a plane, audience members in the theatre, or students in the classroom—are relatively important components of many such jobs."[25] While occupational segregation is partly the consequence of discriminatory decisions made by employers, this is something different, a supply-side factor, where people seek to match their abilities and characteristics to the careers which best suit them—with one particular ability, that of social perceptiveness, being significant.

Other research exploring occupational segregation for this community has identified the role of gender stereotypes. As we have seen, gay men are

frequently stereotyped as feminine and lesbian women are commonly stereotyped as masculine. Extending this, gay men are commonly associated
with female-dominated occupations such as nursing, hairdressing, dancing
and acting, and lesbians with male-dominated occupations such as athletics,
mechanics and truck driving.[26]

The lack of more specific information on this topic led Ashley to undertake a PhD research project exploring this phenomenon further. We already
knew that stereotypes about occupations influence individuals' perceptions
of their own capability, which can lead to decisions to pursue or avoid career
paths based on perceived "fit." We also knew that stereotypes based on gender play an important role in people's identification of suitable career paths.
It had therefore been suggested that LGBTQ+ individuals discard particular
occupations as a result of perceived role incongruity or a "lack of fit."

In order to shed light on whether this internalisation of stereotypes does
indeed drive occupational segregation, Ashley interviewed 40 LGBTQ+ professionals across public and private sector organisations, seeking to understand the impact of internalised stereotypes on their career experiences with
a focus on perceptions of career barriers throughout their vocational tra-
jectories.[27] She found that perceived incongruence between personal and
occupational stereotypes did impact career decisions; however, this impact
depended on the nature of the stereotypes that had been internalised. For
lesbians, internalised stereotypes were focused on presenting and appearing
as less feminine than heterosexual women, leading to pressure to conform
to such appearance in order to be successful in a professional work environment. Gay men on the other hand had internalised stereotypes regarding their behaviour which led to the perception of various career barriers
including not fitting into masculine work environments, risk of bullying and
discrimination, and not being judged on their abilities. Internalised stereotypes therefore impacted the way that lesbians and gay men navigated their
careers.

Unlike the majority of other research undertaken in this area, Ashley also
extended the discussion to other members of the LGBTQ+ community. In
the study, bisexual participants reported feeling that they needed to craft
their careers towards work environments where colleagues were more likely
to hold liberal attitudes about bisexuality. This was therefore less to do with
gendered segregation and more to do with the attitudes of others, akin to the
safe haven hypothesis. Finally transgender participants reported that stereotypes regarding gender non-conformity led to abandonment of career aspirations and encouraged decisions to pursue careers in occupations which were
not prominently gender stereotyped.

An additional consideration for LGBTQ+ individuals in seeking employment is the possibility of needing to travel to different regions for work.
This presents a risk of having to visit conservative countries or areas where
there are prohibitive laws and negative societal attitudes regarding LGBTQ+

identity which could present a risk to their safety. This consideration will be taken into account during the process of career planning and job seeking, potentially imposing a limit on opportunities for individuals within the community.

Recruitment and Selection

In 1968 Philip Goldberg presented students with a number of articles to assess. All students received the same articles but for half of the sample the author was stated as John T. McKay and for the others the *h* in John was swapped for an *a*. This small change led to poorer evaluations of the article's quality.[28]

This ingenious format for research has been used innumerable times since and is known as the Goldberg paradigm. It has been most commonly used in selection studies where the names on CVs or résumés are easily altered. This is not possible with LGBTQ+ candidates so the application needs to be altered in other ways. Usually this is done by sharing some information on the CV which indicates that the applicant is either from the LGBTQ+ community or can be considered to be an ally. This might include indicating involvement in and support of the LGBTQ+ community such as volunteering at a gay-affirming church or being a mentor to LGBTQ+ students. It was found that the LGBTQ+ allies were significantly less likely to be hired. They were also viewed as being less likely to fit in and more likely to experience bullying and harassment in the workplace. Even objective information was rated lower with the LGBTQ+ ally's qualifications being seen as poorer, even though they were the same as those on the non-ally CVs. It was also found that African American hirers were harsher in their evaluations of the LGBTQ+ allies.[29]

Andras Tilcsik employed this particular method in a 2011 study.[30] He identified almost 1,800 job adverts across seven U.S. states and submitted two fake résumés for each opening. Within each pair the applicant designed to be gay listed their experience in a gay student organisation at university while the applicant designed to be heterosexual listed experience as a treasurer of a small left-wing campus organisation. Of the heterosexual applicants 11.5% were invited to interview compared to 7.2% of equally qualified applicants who had worked for a gay organisation. This represents a 40% decrease in the likelihood to be interviewed for gay applicants compared to heterosexual applicants.

Other studies have identified this pattern of bias towards lesbian[31] and transgender job applicants[32] and have demonstrated the effect across different cultural contexts including Greece,[33] Sweden[34] and the UK.[35] Tilcsik's study further revealed that gay applicants were less likely to be invited to interview for roles that emphasised stereotypically masculine traits such as assertiveness and decisiveness, once again highlighting the role of the Implicit Inversion Theory.[36]

Results consistently show differences between ethnic groups with African-American groups holding more negative attitudes towards the LGBTQ+ community. Pew research indicates that attitudes to same-sex marriage within the African-American community have become more positive (rising from 31% in 2001 to 51% in 2019); the change is smaller than that for the white community (from 34% in 2001 to 62% in 2019).[37]

Despite employees in Europe benefiting from relatively progressive protective legislation, research in the last decade has shown that around one in five LGBT employees experience discrimination during recruitment processes and at work, with transgender people coming off the worst.[38]

In a U.S. survey of 6,450 transgender respondents conducted in 2011, 47% reported that they had experienced an adverse job outcome such as being fired, not hired or denied a promotion because of being transgender or gender non-conforming.[39] In Ireland researchers reported that 14% of the transgender respondents believed they had been denied a job on the basis of their transgender identity. Twenty-four per cent of respondents were unemployed and seeking work.[40] A 2014 EU-wide survey of 6,579 transgender respondents found that 37% felt discriminated against because of being transgender when looking for a job.[41]

Wage Inequality

Given all that has been covered, it should be no surprise that another area of inequality relates to earnings. The literature shows that heterosexual men typically earn more than gay men (10–32%)[42] and bisexual men (12%).[43] However, there is typically no difference in the earnings of bisexual and heterosexual women,[44] and lesbians are often reported to earn significantly more than heterosexual women (12%).[45]

In 2007, a Europe-wide review of almost 2,000 transgender people's experiences of inequality and discrimination found that respondents' incomes were disproportionately at the lower end of salary levels.[46] There are, however, variations in income based on gender identity among transgender individuals. Transgender women experienced a reduction of approximately one-third in their earnings after transitioning, whereas transgender men saw an increase in earnings post-transition.[47]

It's important to note that although these findings have been replicated to substantiate the salary gap, each study generates a slightly different figure. The figures are contingent on various factors such as how the salary data is collected and how LGBTQ+ identity is defined. An analysis of the UK Integrated Household Survey found that the difference in wages was only apparent for lesbians and gay men who were partnered compared with heterosexual men and women who had partners. There was in fact no earnings differential for gay men and lesbians who were not partnered.[48] In other words the difference only occurred when the individuals made their identity clear in terms of their relationships.

More recently a study in the United States found that while the immediate earnings gap between LGBTQ+ and heterosexual college graduates was 12%, after a decade this gap almost doubled with LGBTQ+ employees earning 22% less than their heterosexual counterparts.[49] This demonstrates that while some of the earnings differential could be attributed to experiences and decisions before entering a work environment, the gap is further exacerbated by experiences at work. Another factor that could be at play here is the high concentration of LGBTQ+ individuals in stereotypically feminine roles and industries, which are usually lower paid than stereotypically masculine roles and industries.

Promotion Discrimination

The Corporate Closet, written in 1990 by former advertising creative turned academic James D Woods, was a ground-breaking book examining the lives of gay professionals and executives. The book is full of stories of gay men keeping their private and public lives completely separate; not socialising with colleagues after work so that they didn't reveal inadvertently their sexuality; not being able to form strong connections with mentors and then consequently missing out on advice and guidance; not being able to care for colleagues, family and partners who were ill; and being unable to grieve for people who had passed away during the AIDS epidemic of the 1980s. The book gave very clear evidence of how gay men had to watch their own behaviour on a daily basis to ensure that they did not give any indication that they were anything other than heterosexual. There are many case studies in the book including this one:

> Martin thinks that by staying in the closet, he has largely avoided discrimination. He recognises that his sexuality has influenced many of his choices, stunted certain professional relationships, even drawn him to a particular city and line of work.[50]

From another book, here is John describing his experiences:

> I came out after I had had a successful executive career and after I had made contacts across industries from oil and gas to finance over four decades. Before I came out, I made it to the top of [the organisation] without the stigma of being perceived as an outsider. A young person who comes out at the start of his or her career might be denied the opportunities that I was given.

This is John, now Lord, Browne, the former Chief Executive Officer of BP, who kept his sexual orientation a secret throughout whole of his career. In a searingly honest account of his life and career, the word "fear," and those related to it such as fearful, is used nearly 50 times in a book of 200 pages.

It demonstrates that even the most powerful leaders in the world can experience very dark feelings when it comes to being true to themselves and sharing their identity with others.[51]

LGBTQ+ discrimination also occurs in the promotion process. The *LGBT in Britain* report[52] revealed that one in ten employees said they were denied a promotion at work in the previous year because of their LGBTQ+ identity. When further broken down by sexuality and gender identity, this number rises to 24% of transgender respondents compared to 7% of LGBTQ+ people who are not transgender.

This phenomenon has been investigated over the last two decades and has been labelled the gay glass ceiling.[53] The first large-scale study of the effect in the UK found that gay men were significantly less likely to occupy the highest-level managerial positions compared to heterosexual men. The researchers also found that the gay glass ceiling effect was more pronounced for racial minorities.[54]

One of the contributing factors to this phenomenon may be the conflation of gay male identities with typically feminine attributes as described by the Implicit Inversion Theory and Think Manager, Think Heterosexual Male paradigms described in the previous chapter. Gay men are rated as being more communal and suited for "feminine managerial positions" than heterosexual men though less suited for "masculine leadership positions."[55] Stereotypically feminine jobs are generally lower status and lower paid than masculine jobs.

A UK study found that gay men were significantly more likely to be low-level managers and less likely to be in the highest-level management positions compared to heterosexual men.[56] OECD data indicates that the high-level managerial occupation gap amounts to minus 6% for lesbian and bisexual women, and minus 16% for gay and bisexual men.[57]

It may be no surprise to find that those with higher levels of homonegativity also rated gay males as less effective leaders.[58] This gives some insight into the factors that may lead to such discrimination, but research has largely overlooked bisexual and transgender participants.

It also seems that cues such as voice can themselves lead individuals to make biased decisions.[59]

Career Capital

Career capital refers to assets that contribute to the success of an individual's employment and overall career[60] and is divided into three categories:

Knowing-why—the sense of purpose and motivation one has for one's career.
Knowing-how—the skills and knowledge one has.
Knowing-whom—reputation, relationships and networks.

Transgender individuals who transition during their career may experience a negative impact on their career capital. For individuals who change their gender expression (including their name), skills and knowledge acquired before transition may be overlooked or misjudged by new employers.[61] People who have transitioned may also choose to remove themselves from networks that they were part of before transitioning and instead start afresh in a new role, building new relationships without disclosing their transgender identity. This may cancel the knowing-whom capital that an individual has built throughout their career. For example, in one study 7% of respondents said that they had not provided references from a previous job due to their gender history.[62]

Bullying

"Fitting in"—or failing to fit in—has always been an acceptable way of discriminating against people and has long been seen as good management practice. Chester L. Bernard, one of the early management gurus, wrote in his classic work of 1938, *The Functions of the Executive*, that people may not be suitable for an executive role not because of capability but due to the fact that they just didn't fit in.

> Perhaps often, and certainly occasionally, men cannot be promoted or selected, or even must be relieved, because they cannot function because they "do not fit" where there is no question of formal competence. The question of "fitness" involves such matters as education, experience, age, sex, personal distinctions, prestige, race.[63]

Of course, we wouldn't say these things so explicitly today—but at some level do we still think this?

Someone's difference from others can also create a reason for bullying and harassing them. Research conducted with a representative sample of LGBTQ+ employees in the UK found that rates of bullying and harassment for LGBTQ+ staff were two to three times higher than for their straight colleagues: 6.4% of heterosexual respondents experienced bullying and harassment compared to 13.7% of gay men, 16.9% of lesbians and 19.2% of bisexuals. "Having inadequate resources," "not having enough time to carry out the job" and "I cannot follow best practice in the time available" were factors related to bullying, all of which increase stress and lead to poorer performance. At work, LGB individuals would be more likely to:

- receive hostile reactions if they discuss their personal life;
- be insulted or have offensive remarks made about them;
- receive threats from people at work;
- be shouted at;
- experience physical violence at work such as being hit, kicked or pushed around.

Respondents identified the behaviour they found most difficult to deal with at work as being talked to in an insulting or derogatory manner.[64]

The *LGBT in Britain* report[65] shows that almost one in five (18%) of LGBT respondents had been the target of negative comments or behaviour in the previous year from work colleagues because of their LGBT identity. One in eight (12%) transgender respondents had been physically attacked by customers or colleagues in the previous year because of their identity.

Stigmatised individuals can also be excluded or ostracised: not invited to events, not given access to information they need to carry out their work, not communicated with on a regular basis. Here it isn't actions that are done directly to individuals that exclude them from the team but the withdrawal of resources and relationships.

International research also demonstrates that transgender individuals suffer the highest rate of discrimination in employment.[66]

A U.S. report of 2021 found that one in five (20.8%) LGBTQ+ employees reported experiencing physical harassment because of their sexual orientation or gender identity. Instances of physical harassment in the workplace included being punched, hit and beaten up.[67]

Sexual Harassment

Alongside the data about LGBTQ+ workplace bullying, there is an alarming frequency of LGBTQ+ employees' exposure to sexual harassment at work.

In 2019 the TUC conducted a study to investigate the experiences of 1,000 LGBT people at work in the UK. This study showed that 68% of LGBT employees had experienced at least one type of sexual harassment at work and 12% of LGBT women reported being seriously sexually assaulted at work.[68] The bullying behaviours detailed by which LGB respondents found most difficult to deal with included being confronted with unwanted jokes or remarks which have a sexual undertone and being asked intrusive questions about their personal or private life.[69]

A U.S. report of 2021 found that one in four (25.9%) LGBT employees reported experiencing sexual harassment in the workplace because of their sexual orientation and gender identity at some point in their careers. Transgender individuals were twice as likely to report recent experiences of sexual harassment: 22.4% reported sexual harassment in the past five years compared to 11.9% of LGB employees.[70]

In its study the TUC referred to sexual harassment as "a hidden problem"[71] with 65% of those in their report who were harassed not reporting it to their employer. The most common reasons for not reporting were: fear of a negative impact on relationships at work (57%), fear of a negative career impact (44%), belief that the person responsible would not be punished (40%) and requirement to disclose sexual and/or gender identity (25%).

Exclusion and Micro-incivilities

Discrimination and exclusion are not always obvious, even to the perpetrators. The smaller behaviours that can make people feel excluded are most commonly referred to as micro-aggressions—an inadequate term for a number of reasons but most importantly because it eliminates the possibility that behaviour has been carried out unknowingly.

We have met many people who suggest that bias and discrimination can never be carried out unconsciously. Part of the reason people say this is that we imagine three distinct groups of people: victims, perpetrators and bystanders. We can appreciate how we could be victims of such behaviour and will almost certainly have experiences as bystanders, but we can never imagine ourselves as perpetrators. In fact, all of us can find ourselves in each of those three roles. If we believe that such behaviour is always intentional, then when we are accused of being a perpetrator, we should immediately admit to doing so consciously.

The starting point for discussing this topic is often Derald Wing Sue's book *Micro Aggressions in Everyday Life*, which popularised the term. We find his analysis provocative, at times illuminating but also on occasion unhelpful. He identifies several categories of behaviour related to sexual orientation, but they are a mix of overt and subtle categories. The genuinely "micro" behaviours include language which makes assumptions about others, such as girlfriend or boyfriend, wife or husband, and use of terms such as "that is so gay" to identify something as inappropriate or wrong. Other examples could never be considered as anything other than overtly prejudiced, including drawing attention to the sexual behaviour of LGBTQ+ people, abusive language and viewing heterosexual relationships normal while anything else is either abnormal or sinful. He even has a separate category for homophobia, which includes minimising contact with LGBTQ+ individuals and "washing one's hands immediately after shaking hands with a gay man [and] making sure that sons and daughters are restricted from going to a gay neighbour's home."[72]

We prefer the term micro-incivilities which we define as

The kinds of daily, commonplace behaviors or aspects of an environment which signal, wittingly or unwittingly, to members of out-groups that they do not belong and are not welcome.[73]

This is extending the work on organisational citizenship behaviours,[74] which are discretionary and which people engage in in order to create a positive and inclusive working environment. Incivilities have been recognised as counter-productive behaviours which create climates of mistrust, suspicion and disengagement. This is a useful concept which has been applied to organisations for several decades, and there is a research base which demonstrates the value of civil behaviours and the corrosiveness in incivilities.

A literature review[75] by Kevin Nadal and his colleagues provides more guidance on micro-incivilities, helpfully broken down by identity. Micro-incivilities experienced by lesbians include families and friends failing to listen to and understand the person by saying for example that they are "just going through a phase"; being expected to dress consistently with other women (gender conformity), and lesbian and bisexual women being seen as more exotic and then being inappropriately propositioned by men.

Gay men experienced derogatory heterosexist language such as being called gay, faggot or fag. Phrases such as "that's so gay" will be used more often by individuals if others in the group use them frequently. Such expressions are also most likely to be used by those who find it difficult to deal with people who do not conform to their picture of masculinity; people using such language may be revealing their own discomfort. Stereotypes about gay men and assumptions about their character and behaviour are, not surprisingly, another category of micro-incivility. These stereotypes include being fun, effeminate, fashionable dressers, sexually predatory and promiscuous.

There are several categories of micro-incivilities relating specifically to bisexual people. These include hostility from both LGBTQ+ and straight people; denial and dismissal of a person's bisexual identity; unintelligibility—that is, people being unable to understand someone who may be attracted to both sexes; and people feeling a pressure to re-label their identity, so it best matches the relationship they are in at that time (gay, lesbian or heterosexual). When interacting with other members of the LGBTQ+ community, bisexual people felt they were struggling to be accepted and were looked down on. They experienced dating exclusion, where women would not go out with bisexual women because of their sexual identity and being stereotyped as hypersexual and therefore unable to be monogamous.

It has been known for some time that bisexuals experience different stressors than lesbians and gay men. In a provocative paper written over 40 years ago, researcher A. P. McDonald Jr pointed out that researchers have treated bisexuals as if they were lesbians or gay, despite some of the research into gay and lesbian experience having in their samples anywhere between a quarter and over a half stating that they were bisexual. This feeds into the narrative that bisexuals are really lesbian or gay but won't admit to it.[76]

As in other research, the use of humour to cover denigrating comments was ultimately the most difficult type of behaviour to respond to. The other key difficulty is shared by everyone with an invisible identity: do you accept others' denigratory behaviour towards people with whom you share an identity, or do you challenge it? Confronting such behaviour carries the risk of revealing your hidden identity to others.[77]

Although these behaviours may seem small, they can contribute to or cause the discriminatory outcomes we have explored. This is evident in one study[78] where a group of participants applied for jobs in stores. Although there was no evidence that gay and lesbian participants were formally discriminated against, the researchers did identify subtle differences in treatment such as fewer conversational exchanges and shorter interactions with gay and lesbian applicants compared to heterosexual applicants. Such subtle forms of discrimination can be just as impactful and can indeed lead to the differences in opportunities and success rates.

The actions also detrimentally impact the mental health of members of marginalised social groups,[79] with studies correlating micro-incivilities with lower levels of self-esteem, higher prevalence of depressive symptoms lower levels of psychological well-being, higher prevalence of binge drinking and higher negative emotional intensity.

Each of us can reflect on our own experiences and interactions. The instances where we have unconsciously made heteronormative assumptions by, for example, assuming that someone's romantic partner is of the opposite gender, or maybe making an assumption that your male colleague who is about to take parental leave will be back in a week or two while his wife stays at home with the child, without ever having been informed of his sexuality or heard him talk about his partner.

An LGBTQ+ person on the receiving end of these assumptions faces a deeply unpleasant situation. Some LGBTQ+ colleagues may choose to correct the assumption in the moment while others may decide not to. In either case this kind of assumption serves as a reminder of the deeply ingrained default identity in our society and how this can make others feel.

While most people align across their birth-assigned sex, gender identity, gender expression and how everyone else interprets their gender, transgender and non-binary people may, as part of their transition or gender affirmation, ask to be referred to using pronouns that differ from those that are commonly assumed or those that were previously being used. For example, somebody who was assigned the sex male at birth but who identifies as female may change their name and ask to be referred to using the pronouns she and her. Somebody who does not identify as male or female may ask to be referred to as they and them.

Misgendering occurs when we refer to a person using incorrect pronouns, thereby applying a gender with which they don't identify. Using incorrect pronouns to refer to somebody may be an automatic action that is not intended to be harmful, but it does have a negative impact on the individual. It leads to feelings of being othered and stigmatised for having an identity that is perceived to be different. This can impact mental health and well-being. It also leads to the individual having to decide on whether to correct the person or minimise the likelihood of an uncomfortable discussion by not saying anything.

Key Points

The stereotyping and stigmatisation that were described in Chapter 4 have a significant impact in terms of discrimination that is experienced in employment.

The impact of stigmatisation can be seen before people join an organisation and before they become adults. For LGBTQ+ individuals, it happens when they are younger and when they start looking at the types of careers and jobs where they feel they will be accepted and valued. This has the impact of limiting their career choices and leads to occupational segregation.

Not surprisingly, given the negative evaluations that are made of LGBTQ+ members, discrimination occurs from the recruitment and selection stage, through to promotion. Knowing someone's sexual orientation has an impact on how they are viewed. Over the course of a career this has a considerable impact on pay—and there is a penalty for sharing your identity in many organisations.

People in the LGBTQ+ community are also more likely to experience bullying and harassment in the workplace, with it being most pronounced for transgender and non-binary individuals. This creates unwelcome and stressful working environments for individuals which has an impact on their performance, which then will impact their performance evaluations.

The behaviours that people are subject to aren't necessarily overt and blatant. Many of the behaviours may in fact be subtle, indirect and not necessarily consciously carried out in order to exclude someone. The perpetrators may not be aware of having done anything wrong, others who are present may not be aware of anything untoward, but those who are impacted by the behaviour will definitely feel uncomfortable and excluded.

The subtlety of the behaviour, however, means that there will be a degree of ambiguity. Individuals will be left wondering whether the slight was intended and indeed questioning whether it was offensive at all. When this happens on a regular basis, the questioning of what happened, and the rumination that accompanies it, creates longer-term stress for individuals.

Finally, it's also important to be aware that the micro-incivilities that people experience will be different according to the person's identity. Being aware of the types of behaviour that can cause exclusion and stress means that we can all watch out for them at work and give feedback to those engaging in such actions in appropriate way. Listening to the experiences of LGBTQ+ colleagues therefore is an essential starting point for creating a more inclusive workplace.

Notes

1 Bos, A.E.R., Pryor, J.B., Reeder, G.D., & Stutterheim, S.E. (2013). Stigma: Advances in Theory and Research. *Basic and Applied Social Psychology*, 35(1), pp. 1–9. https://doi.org/10.1080/01973533.2012.746147
2 Hetherington, C. (1991). Life Planning and Career Counseling with Gay and Lesbian Students. In: *Beyond Tolerance: Gays, Lesbians, and Bisexuals on Campus*, pp. 131–145. Routledge.
3 Schmidt, C.K., & Nilsson, J.E. (2006). The Effects of Simultaneous Developmental Processes: Factors Relating to the Career Development of Lesbian, Gay and Bisexual Youth. *The Career Development Quarterly*, 55(1), pp. 22–37.

4 Kosciw, J.G., Gretak, E.A., Palmer, N.A., & Boesen, M.J. (2014). *The 2013 National School Climate Survey: The Experiences of Lesbian, Gay, Bisexual and Transgender Youth in Our Nation's Schools*. Gay, Lesbian and Straight Education Network (GLSEN). New York, NY.

5 Allen, D.J., & Oleson, T. (1999). Shame and Internalized Homophobia in Gay Men. *Journal of Homosexuality*, 37(3), pp. 33–43. https://doi.org/10.1300/J082v37n03_03

6 Szymanski, D.M., Chung, Y.B., & Balsam, K.F. (2001). Psychosocial Correlates of Internalized Homophobia in Lesbians. *Measurement and Evaluation in Counseling and Development*, 34(1), pp. 27–38.

7 Wada, K., Mcgroarty, E.J., Tomaro, J., & Amundsen-Dainow, E. (2019). Affirmative Career Counselling with Transgender and Gender Nonconforming Clients: A Social Justice Perspective. *Canadian Journal of Counselling and Psychotherapy*, 53(3), pp. 255–275.

8 Effrig, J.C., Bieschke, K.J., & Locke, B.D. (2011). Examining Victimization and Psychological Distress in Transgender College Students. *Journal of College Counseling*, 14(2), pp. 143–157.

9 Scott, D.A., Belke, S.L., & Barfield, H.G. (2011). Career Development with Transgender College Students: Implications for Career and Employment Counselors. *Journal of Employment Counseling*, 48(3), pp. 105–113. https://doi.org/10.1002/j.2161-1920.2011.tb01116.x

10 Tilcsik, A., Anteby, M., & Knight, C.R. (2015). Concealable Stigma and Occupational Segregation: Toward a Theory of Gay and Lesbian Occupations. *Administrative Science Quarterly*, 60(3), pp. 446–481. Crossref. ISI.

11 Ellis, H. (1915). *Sexual Inversion* (3rd ed.). London: Wilson and Macmillan, p. 194.

12 Tilcsik, A., Anteby, M., & Knight, C.R. (2015). Concealable Stigma and Occupational Segregation: Toward a Theory of Gay and Lesbian Occupations. *Administrative Science Quarterly*, 60(3), pp. 446–481.

13 Tilcsik, A., Anteby, M., & Knight, C.R. (2015). Concealable Stigma and Occupational Segregation: Toward a Theory of Gay and Lesbian Occupations. *Administrative Science Quarterly*, 60(3), pp. 446–481.

14 Ueno, K., Roach, T., & Peña-Talamantes, A.E. (2013). Sexual Orientation and Gender Typicality of the Occupation in Young Adulthood. *Social Forces*, 92(1), pp. 81–108. https://doi.org/10.1093/sf/sot067

15 Pichler, S., & Ruggs, E.N. (2015). LGBT Workers. In: Adrienne J. Colella, & Eden B. King (eds.), *The Oxford Handbook of Workplace Discrimination*. Oxford Library of Psychology, pp. 665–701.

16 Ragins, B.R. (2004). Sexual Orientation in the Workplace: The Unique Work and Career Experiences of Gay, Lesbian and Bisexual Workers. In: *Research in Personnel and Human Resources Management*. New York, NY: JAI Press, pp. 35–129.

17 Chung, Y.B., Williams, W., & Dispenza, F. (2009). Validating Work Discrimination and Coping Strategy Models for Sexual Minorities. *The Career Development Quarterly*, 58(2), pp. 162–170.

18 Schneider, M.S., & Dimito, A. (2010). Factors Influencing the Career and Academic Choices of Lesbian, Gay, Bisexual, and Transgender People. *Journal of Homosexuality*, 57(10), pp. 1355–1369.

19 Fielden, S.L., & Jepson, H. (2018). Lesbian Career Experiences. In: *Research Handbook of Diversity and Careers*. Edward Elgar Publishing.

20 Colgan, F. (2018). Coming Out of the Closet? The Implications of Increasing Visibility and Voice for the Career Development of LGB Employees in UK Private Sector Organisations. In: *Research Handbook of Diversity and Careers*. Edward Elgar Publishing.

21 Marlow, S., Greene, F.J., & Coad, A. (2018). Advancing Gendered Analyses of Entrepreneurship: A Critical Exploration of Entrepreneurial Activity Among Gay

Men and Lesbian Women. *British Journal of Management*, 29(1), pp. 118–135. https://doi.org/10.1111/1467-8551.12221

22 Marlow, S., Greene, F.J., & Coad, A. (2018). Advancing Gendered Analyses of Entrepreneurship: A Critical Exploration of Entrepreneurial Activity Among Gay Men and Lesbian Women. *British Journal of Management*, 29(1), pp. 118–135. https://doi.org/10.1111/1467-8551.12221

23 Ng, E.S., Schweitzer, L., & Lyons, S.T. (2012). Anticipated Discrimination and a Career Choice in Nonprofit A Study of Early Career Lesbian, Gay, Bisexual, Transgendered (LGBT) Job Seekers. *Review of Public Personnel Administration*, 32(4), pp. 332–352.

24 Tilcsik, A., Anteby, M., & Knight, C.R. (2015). Concealable Stigma and Occupational Segregation: Toward a Theory of Gay and Lesbian Occupations. *Administrative Science Quarterly*, 60(3), pp. 446–481.

25 Tilcsik, A., Anteby, M., & Knight, C.R. (2015). Concealable Stigma and Occupational Segregation: Toward a Theory of Gay and Lesbian Occupations. *Administrative Science Quarterly*, 60(3), pp. 446–481.

26 Rule, N.O., Bjornsdottir, R.T., Tskhay, K.O., & Ambady, N. (2016). Subtle Perceptions of Male Sexual Orientation Influence Occupational Opportunities. *Journal of Applied Psychology*, 101(12), pp. 1687–1704.

27 Williams, A. (2021). *How Do Stereotypes of Sexuality and Gender Influence LGBT Career Construction?* (PhD Thesis).

28 Goldberg, P. (1968). Are Women Prejudiced Against Women? *Transaction*, 5, pp. 28–30.

29 LeCroy, V.R., & Rodefer, J.S. (2019). The Influence of Job Candidate LGBT Association on Hiring Decisions. *North American Journal of Psychology*, 21(2), pp. 373–385. https://go.openathens.net/redirector/leeds.ac.uk?url=www.proquest.com/scholarly-journals/influence-job-candidate-lgbt-association-on/docview/2236686217/se-2

30 Tilcsik, A. (2011). Pride and Prejudice: Employment Discrimination Against Openly Gay Men in the United States. *The American Journal of Sociology*, 117(2), pp. 586–626.

31 Weichselbaumer, D. (2003). Sexual Orientation Discrimination in Hiring. *Labour Economics*, 10(6), pp. 629–642. https://doi.org/10.1016/S0927-5371(03)00074-5

32 Dispenza, F., Watson, L.B., Chung, Y.B., & Brack, G. (2012). Experience of Career-Related Discrimination for Female-to-Male Transgender Persons: A Qualitative Study. *The Career Development Quarterly*, 60, pp. 65–81.

33 Drydakis, N. (2009). Sexual Orientation Discrimination in the Labour Market. *Labour Economics*, 16(4), pp. 364–372.

34 Ahmed, A.M., Andersson, L., & Hammarstedt, M. (2013). Are Gay Men and Lesbians Discriminated Against in the Hiring Process? *Southern Economic Journal*, 79(3), pp. 565–585.

35 Drydakis, N. (2014). Sexual Orientation Discrimination in the United Kingdom's Labour Market: A Field Experiment (No. 8741).

36 Tilcsik, A. (2011). Pride and Prejudice: Employment Discrimination Against Openly Gay Men in the United States. *The American Journal of Sociology*, 117(2), pp. 586–626.

37 Pew Research Center. (2019, May 14). www.pewresearch.org/religion/fact-sheet/changing-attitudes-on-gay-marriage/

38 Bachmann, C.L., & Gooch, B. (2018). *LGBT in Britain: Work Report*. Stonewall.

39 Grant, J.M., Mottet, L.A., Tanis, J., Harrison, J., Herman, J.L., & Keisling, M. (2011). *Injustice at Every Turn: A Report of the National Transgender Discrimination Survey*. Washington: National Center for Transgender Equality and National Gay and Lesbian Task Force.

40 McNeil, J., Bailey, L., Ellis, S., & Regan, M. (2013). *Speaking from the Margins: Trans Mental Health and Wellbeing in Ireland.* Dublin: Transgender Equality Network Ireland.

41 https://fra.europa.eu/en/publication/2014/being-trans-eu-comparative-analysis-eu-lgbt-survey-data

42 Badgett, M.V., Lau, H., Sears, B., & Ho, D. (2007). *Bias in the Workplace: Consistent Evidence of Sexual Orientation and Gender Identity Discrimination.* Williams Institute, University of California School of Law.

43 Uhrig, N.S.C. (2015). Sexual Orientation and Poverty in the UK: A Review and Top-Line Findings from the UK Household Longitudinal Study. *Journal of Research in Gender Studies*, 5(1), pp. 23–72.

44 Uhrig, N.S.C. (2015). Sexual Orientation and Poverty in the UK: A Review and Top-Line Findings from the UK Household Longitudinal Study. *Journal of Research in Gender Studies*, 5(1), pp. 23–72.

45 Brewster, M.E. (2017). Lesbian Women and Household Labor Division: A Systematic Review of Scholarly Research from 2000 to 2015. *Journal of Lesbian Studies*, 21(1), pp. 47–69. https://doi.org/10.1080/10894160.2016.1142350

46 Whittle, S., Turner, L., Coombs, R., & Rhodes, S. (2008). *Transgender Eurostudy: Legal Survey and Focus on the Transgender Experience of Health Care.* Brussels: ILGA Europe.

47 Schilt, K., & Wiswall, M. (2008). Before and After: Gender Transitions, Human Capital, and Workplace Experiences. *The BE Journal of Economic Analysis & Policy*, 8(1).

48 Aksoy, C.G., Carpenter, C.S., & Frank, J. (2016). Sexual Orientation and Earnings: New Evidence from the UK. Working Paper No. 196, European Bank for Reconstruction and Development.

49 Folch, M. (2022, April 1). *The LGBTQ+ Gap: Recent Estimates for Young Adults in the United States.* http://dx.doi.org/10.2139/ssrn.4072893

50 Woods, J.D., & Lucas, J.H. (1993). *The Corporate Closet: The Professional Lives of Gay Men in America/James D. Woods with Jay H. Lucas.* The Free Press, p. 11.

51 Browne, J. (2014). *The Glass Closet: Why Coming out is Good Business.* Random House.

52 Bachmann, C.L., & Gooch, B. (2018). *LGBT in Britain: Work Report.* Stonewall.

53 Frank, J. (2006). Gay Glass Ceilings. *Economica*, 73(291), pp. 485–508.

54 Carpenter, C.S., Frank, J., Aksoy, C.G., & Huffman, M.L. (2018). *Gay Glass Ceilings: Sexual Orientation and Workplace Authority in the UK, No 11574, IZA Discussion Papers.* Institute of Labor Economics (IZA). https://EconPapers.repec.org/RePEc:iza:izadps:dp11574

55 Barrantes, R.J., & Eaton, A.A. (2018). Sexual Orientation and Leadership Suitability: How Being a Gay Man Affects Perceptions of Fit in Gender-Stereotyped Positions. *Sex Roles*, pp. 1–16.

56 Aksoy, C.G., Carpenter, C.S., Frank, J., Huffman, M.L. (2019). Gay Glass Ceilings: Sexual Orientation and Workplace Authority in the UK. *Journal of Economic Behavior & Organization*, 159, pp. 167–180. https://doi.org/10.1016/j.jebo.2019.01.013

57 OECD. (2019). *Society at a Glance 2019: OECD Social Indicators.* Paris: OECD Publishing. https://doi.org/10.1787/soc_glance-2019-en

58 Morton, J.W. (2017). Think Leader, Think Heterosexual Male? The Perceived Leadership Effectiveness of Gay Male Leaders. *Canadian Journal of Administrative Sciences*, 34(2), pp. 159–169. https://doi.org/10.1002/cjas.1434

59 Fasoli, F., Hegarty, P., & Frost, D.M. (2021). Stigmatization of 'Gay-Sounding' Voices: The Role of Heterosexual, Lesbian, and Gay Individuals' Essentialist Beliefs. *British Journal of Social Psychology*, 60, pp. 826–850. https://doi.org/10.1111/bjso.12442

60 Inkson, K., & Arthur, M.B. (2001). How to be a Successful Career Capitalist. *Organizational Dynamics*, 30(1), pp. 48–61. https://doi.org/10.1016/S0090-2616(01)00040-7

61 Sangganjanavanich, V.F. (2009). Career Development Practitioners as Advocates for Transgender Individuals: Understanding Gender Transition. *Journal of Employment Counseling*, 46(3), pp. 128–135. https://doi.org/10.1002/j.2161-1920.2009.tb00075.x

62 McNeil, J., Bailey, L., Ellis, S., & Regan, M. (2013). *Speaking from the Margins: Trans Mental Health and Wellbeing in Ireland*. Dublin: Transgender Equality Network Ireland.

63 Barnard, C.I. (1938). *The Functions of the Executive*. Harvard University Press, p. 224.

64 Hoel, H., Lewis, D., & Einarsdóttir, A. (2014). *The Ups and Downs of LGBs' Workplace Experiences*. Manchester Business School.

65 Bachmann, C.L., & Gooch, B. (2018). *LGBT in Britain: Work Report*. Stonewall.

66 ILO. (2013). *Discrimination at Work on the Basis of Sexual Orientation and Gender Identity: Results of Pilot Research*. www.ilo.org/gb/GBSessions/GB319/lils/WCMS_221728/lang—en/index.htm

67 Sears, B., Mallory, C., Flores, A.R., & Conron, K.J. (2021). *LGBT People's Experiences of Workplace Discrimination and Harassment*. William Institute UCLA School of Law.

68 TUC Sexual Harassment Report. www.tuc.org.uk/sites/default/files/LGBT_Sexual_Harassment_Report_0.pdf

69 Hoel, H., Lewis, D., & Einarsdóttir, A. (2014). *The Ups and Downs of LGBs' Workplace Experiences*. Manchester Business School.

70 Sears, B., Mallory, C., Flores, A.R., & Conron, K.J. (2021). *LGBT People's Experiences of Workplace Discrimination and Harassment*. William Institute UCLA School of Law.

71 TUC Sexual Harassment Report. www.tuc.org.uk/sites/default/files/LGBT_Sexual_Harassment_Report_0.pdf, p. 4.

72 Sue, D.W. (2010). *Microaggressions in Everyday Life: Race, Gender, and Sexual Orientation*. Hoboken, NJ: Wiley, p. 193.

73 Kandola, B. (2018). *Racism at Work: The Danger of Indifference*. Kidlington: Pearn Kandola Publishing.

74 Borman, W.C., & Motowidlo, S.J. (1993). Expanding the Criterion Domain to Include Elements of Contextual Performance. In: N. Schmitt, & W.C. Borman (eds.), *Personnel Selection*. San Francisco: Jossey-Bass, p. 73.

75 Nadal, K.L., Whitman, C.N., Davis, L.S., Erazo, T., & Davidoff, K.C. (2016). Microaggressions Toward Lesbian, Gay, Bisexual, Transgender, Queer, and Genderqueer People: A Review of the Literature. *The Journal of Sex Research*, 53, pp. 4–5, pp. 488–508. https://doi.org/10.1080/00224499.2016.1142495

76 MacDonald, A.P. (1981). Bisexuality: Some Comments on Research and Theory. *Journal of Homosexuality*, 6(3), pp. 21–36.

77 Platt, L.F., & Lenzen, A.L. (2013). Sexual Orientation Microaggressions and the Experience of Sexual Minorities. *Journal of Homosexuality*, 60(7), pp. 1011–1034, https://doi.org/10.1080/00918369.2013.774878

78 Hebl, M.R., Foster, J.B., Mannix, L.M., & Dovidio, J.F. (2002). Formal and Interpersonal Discrimination: A Field Study of Bias Toward Homosexual Applicants. *Personality and Social Psychology Bulletin*, 28(6), pp. 815–825.

79 Nadal, K.L., Whitman, C.N., Davis, L.S., Erazo, T., & Davidoff, K.C. (2016). Microaggressions Toward Lesbian, Gay, Bisexual, Transgender, Queer, and Genderqueer People: A Review of the Literature. *The Journal of Sex Research*, 53(4–5), pp. 488–508. https://doi.org/10.1080/00224499.2016.1142495

Chapter 6

The Impact of Exclusion

The discrimination and exclusion faced by the LGBTQ+ community is extensive and impacts all areas of working life including recruitment, development, networking and social interaction. In this chapter we explore the impact of these experiences for LGBTQ+ employees on their well-being, how they manage their identity at work as well as on the impact on organisations.

Impact on Individuals

Exclusion has a considerable impact on individuals, and in this section, we look at unemployment, career aspirations and opportunities as well as identity management.

Unemployment

While overall labour market data on LGBTQ+ people is limited, the research typically shows that lesbian, gay and bisexual workers are more likely to be unemployed than heterosexual individuals and that transgender individuals face higher rates of unemployment again. For example, the EU labour force survey found that people in same-sex partnerships were more likely to be unemployed than individuals in heterosexual partnerships—5.9% versus 4.6%.[1]

In a UK-wide LGBT survey conducted in 2017, 80% of respondents aged 16–64 reported that they had been in employment at some point in the 12 months preceding the survey.[2] For transgender women, this dropped to 65% and for transgender men, 57%. In fact, transgender people have been reported to be more likely to be unemployed compared to any other minority group. A large-scale U.S. study of over 6,000 transgender individuals found they were twice as likely to experience unemployment than the general population.[3]

These findings apply regardless of the commonly reported education gap, which shows that people in a same-sex partnership are on average more highly educated than their heterosexual counterparts (44% compared to

DOI: 10.4324/9781003489580-8

28%).[4] Similarly transgender people in the afore-mentioned U.S. study were reported to have attended college or gained a college degree or higher at 1.74 times the rate of the general population at that time.[5] Although there is no immediately apparent reason for this education gap, some researchers have hypothesised that LGBTQ+ individuals may seek to attain higher levels of education in order to mitigate any potential effect on earnings due to discrimination and to gain access to more tolerant workplaces. Another plausible reason for this finding could be that LGBTQ+ people who are more highly educated may be more likely to disclose their identity in such a study.

While this research cannot draw a causal link between unemployment rates and experiences of discrimination, the studies typically report anecdotal evidence of a connection. In the EU labour force survey, those in same-sex partnerships were much more likely than individuals in a heterosexual partnership to report that they had left their last job because they were dismissed or made redundant (20% versus 13%), because a job of limited duration had ended (26% versus 15%) or for "other reasons" (16% versus 10%). In the UK, almost one in five LGBT people (18%) who were looking for work said they were discriminated against because of their sexual orientation and/ or gender identity while trying to get a job in the previous year.[6] Understandably, workplace bullying reduces job satisfaction for sexual minorities,[7] and in addition, they feel they will have fewer promotion opportunities and are concerned about how others perceive them.

Impact on Career Choice, Aspirations and Opportunities

In the previous chapter we looked at the ways in which stereotypes of and attitudes to the LGBTQ+ community impact workplace experiences through discrimination and exclusion. But the impact of such attitudes extends beyond externally imposed limitations and restrictions to influence how LGBTQ+ people think about themselves and their own abilities which was discussed in Chapter 4.

This internalisation of homophobia and transphobia often impacts career decision-making processes.[8] It has been shown to:

- Impact the process of career exploration for lesbians.[9]
- Result in decreased self-esteem and confidence, which then negatively affects the pursuit of career advancement and opportunities.[10,11]
- Influence a person's self-perception of their own abilities, creating a perception or anticipation of career barriers[12]
- Influence a person's career interests and choices.[13]

To understand the process behind this, researchers have investigated the aspirations of children and how stereotypes impact their thinking. A large

UK study showed that gender stereotyping of occupations was evident from as early as age seven. Overwhelmingly boys showed aspiration for typically masculine roles such as engineering and girls aspired to take on traditionally feminine roles such as teaching.[14]

The inverse of this finding has been identified in the career aspirations of lesbian and gay people. For example gay men, compared to heterosexual men, typically indicate a greater interest for female-typed occupations and lesbian women often indicate a greater interest for male-typed occupations than heterosexual women.[15] Conversely, gay men tend to express less interest in stereotypically masculine careers and lesbians tend to express less interest in stereotypically feminine careers compared to heterosexual workers.[16] This preference for gender atypical work has been shown to be stronger for gay men than for lesbians,[17] which fits with the finding that the inversion of gender stereotyping may be less prevalent for lesbians than for gay men.[18]

Research has also suggested that lesbian women are more likely to report that their sexual orientation has opened up career opportunities for them. The researchers theorised that non-heterosexual identity can negate the strict gendered social expectations of women and create a greater sense of freedom to pursue non-traditional career paths.[19]

Identity Management

The invisible nature of LGBTQ+ identity introduces challenges related to disclosure. If an individual feels that their work environment is not welcoming or safe to disclose LGBTQ+ identity, they may choose instead to conceal their sexual orientation or gender identity. LGB employees who have witnessed—but not been subject to—homophobic incidents are more likely to employ identity management techniques such as concealment.[20] Surveys of UK employees have demonstrated that 20% of LGB, 50% of non-binary and 51% of transgender employees have concealed their sexual or gender identity at work due to concerns regarding discrimination.[21]

Data collected in the United States in 2008 showed that 51% of surveyed LGBT individuals hid their identity from most people at work. In this report only 5% of 18–24-year-olds were completely out at work, compared to more than 20% in older age cohorts.[22] When younger workers were asked why they had not disclosed:

- 65% reported that they didn't want to make others feel uncomfortable
- 39% did not want to risk losing connections with co-workers
- 41% said there was a possibility of being stereotyped
- 28% felt disclosure might be an obstacle to their career advancement or development opportunities
- 13% feared for their personal safety.

Among transgender respondents, 42% hadn't disclosed because they feared being sacked and 40% feared for their personal safety. The survey was repeated a decade later, at which point 46% of LGBTQ+ workers reported that they had not disclosed because:

- 38% said there was a possibility of being stereotyped
- 36% felt it would make others feel uncomfortable
- 31% did not want to risk losing connections with co-workers.

Concealment of identity has many adverse effects on an individual, such as reduced self-confidence, diminished performance, increased isolation and ego depletion.[23] Furthermore employees who hide their identity might forfeit certain employment rewards, such as partner benefits.[24] Research consistently indicates that employees are more open about their identity in supportive work environments.[25]

The decision to disclose one's identity is often referred to in binary terms: to reveal or to conceal? However, the reality of identity disclosure in the workplace is very different. It is not a one-time event but an ongoing process that involves continual negotiation throughout an individual's career. This process includes a cost/benefit analysis considering contextual and personal factors.[26] Identity management can therefore be conceptualised on a continuum ranging from full disclosure to all colleagues to not revealing one's identity to anybody at work. Along this continuum, employees may use specific strategies to manage the (partial) concealment of their identity—for example by disclosing to some but maintaining a false identity with others. This strategy could include avoiding questions about life outside of work or giving false information about their partner's gender. For example, in one U.S. study 23% of gay men and 15% of lesbian women reported that passing as heterosexual had helped their careers. Reported strategies for men included modifying their lifestyle, mannerisms or voice, whereas strategies for women involved changing their appearance, including clothing and hairstyle.[27]

Concealment Behaviours

Concealment behaviours can be divided into three main forms: inhibition, counterfeiting and avoidance.

Inhibition can include people declining to disclose aspects of themselves, even beyond the details of their concealable stigmatised identity. Individuals may also limit their emotional expression and their openness with others, and they can be adept at steering the conversation away from revealing topics such as the topic of partners.[28]

Counterfeiting involves engaging in behaviours in order to avoid stereotypical judgement; for example, gay men may actively participate in stereotypically masculine activities such as sports in order to comply with male gender norms. This behaviour could also include dressing in a particular way to blend in[29] and modifying the tone of their voice or the way that they walk in order to avoid detection of their sexual orientation.[30]

Avoidance of disclosure encapsulates a range of behaviours. Individuals may avoid certain situations where there is an increased risk of victimisation. In a study of gay males from the US and UK, participants were asked to describe retrospectively their self-preservation behaviours in high school:

- 56% reported socially isolating and withdrawing
- 34% feigned an illness
- 56% engaged in truancy.[31]

Furthermore, bisexual people are more likely to conceal their identity at work. Those in same-gender relationships may choose to identify themselves as lesbian or gay to others as this is perceived to have greater legitimacy than a bisexual identity.[32]

The organisational context has a significant impact on whether LGBTQ+ people disclose their identity at work. If people don't feel safe, they will not disclose. Acts of exclusion add up to create a perception that the environment is not safe and welcoming, and that prejudiced attitudes prevail, even if unconsciously.

Personal Impact of Concealment

Concealment of identity is associated with many negative personal outcomes in cognitive energy, wellbeing, self-evaluations and behaviour.

Cognitive Impact

Tell me the first word that comes into your head as long as it's not elephant. The simple exercise reveals how trying to suppress a thought only means that it is more likely to emerge-something known as the rebound effect.[33]

The same thing happens with concealment where individuals are trying to suppress their thoughts about a particular aspect of their identity—which is much easier said than done. It takes mental energy which ultimately results in a cognitive drain.

Researchers identified the impact of this phenomenon by working with women who felt stigmatised by experiences of abortion. In order to conceal this experience, they suppressed thoughts about it. The suppression however only led to intrusive thoughts and then to psychological distress.[34] When attempting to put something out of our minds and protect a secret we become more preoccupied with it and expend greater cognitive resource attending to intrusive thoughts.[35] A constant preoccupation with suppressing thoughts and aspects of identity requires numerous conscious and unconscious processes, which means we have fewer cognitive resources available for other tasks, like doing our job. Self-vigilance will also be needed to monitor one's behaviour and speech. For example, when discussing plans for the weekend an individual concealing their sexual orientation may be preoccupied with self-monitoring to ensure that they do not reveal the gender of their partner during the conversation.

To self-regulate an individual must do two things: first, monitor their behaviour for signs of deviating from a norm and second, alter or override those behavioural impulses. These two functions have different neurological bases, with monitoring processes associated with the anterior cingulate cortex and alteration managed by the prefrontal cortex.[36] During a conversation, concealment requires both of these behaviours with a significant impact on performance. A conversation involving concealment of their sexual orientation lasting just 10 minutes led to worse performance on a spatial ability test that was taken subsequently. A follow-up study showed that simply monitoring one's speech for content to conceal, even when no alteration was required, had an impact on an individual's interpersonal interactions. Specifically, this act of monitoring was sufficient to reduce the politeness with which one responded to an annoying email. The effort involved in the acts of concealment of their identity also led to a depletion of physical strength when compared with a control group. The overall conclusion is that self-regulation to maintain concealment may lead to cognitive depletion over time, which impacts behaviour in other negative ways.[37]

Well-being Impact

Individuals who conceal their identity are more likely to report negative affects,[38] including shame,[39] guilt,[40] frustration[41] and repressed anger.[42] Even the memory of concealing one's identity can trigger such negative affects. A study of lesbian, gay and bisexual individuals in the Netherlands found that after recalling a past experience of concealment participants reported less positive affect than those who had been asked to recall an instance of disclosing one's identity.[43] In another study gay men who concealed their identity reported more depression and poorer overall psychological well-being than those who had disclosed.[44]

A 14-day study with 33 lesbians and 51 gay men found that they reported greater psychological well-being (including self-acceptance, purpose in life, autonomy and mastery) on days when they had disclosed their sexual orientation compared with days when they had concealed. A later study found that on average participants reported more positive feelings, higher self-esteem and more satisfaction with life on days when they disclosed their sexual orientation compared to days when they concealed their sexual orientation. Active suppression of thoughts about sexual orientation predicted lower psychological well-being at the end of each day and also at a two-month follow-up.[45]

In one group of studies, repression and inhibition have been shown to affect immune functions and health outcomes, whereas the expression of emotion (such as writing about a traumatic experience) leads to an improvement in immune functions and decreases negative physical health symptoms.[46] HIV infection has been shown to advance more rapidly among gay men who conceal their identity compared to those who are open about their sexual orientation.[47]

Self-evaluation Impact

Being open about one's sexual orientation in a supportive environment has been associated with lower depression and higher self-esteem.[48] The acts of expressing oneself and one's emotions are important factors in maintaining physical and mental health; however, this protective benefit is denied to LGBTQ+ individuals who conceal their identity.[49] A lack of access to networks and communities of other LGBTQ+ people caused by concealment means that such individuals cannot benefit from the positive self-esteem effects of affiliation with other similarly stigmatised people.[50]

On the other hand, college students with concealable stigmas felt better about themselves when they were around others who were like them as opposed to when they were with those who were not similarly stigmatised.[51]

Behavioural Impact

The avoidant behaviours we have described undoubtedly have implications for an individual's social interactions. Individuals who conceal their identity do not benefit from group affiliation and the benefits that come with it including stress reduction, life satisfaction and support. Being with people like oneself, who have affirmed positive attitudes about the group, can serve a protective function against the stigma that they experience. However, concealment and the avoidance it requires may limit an individual's access to such support. Avoidant behaviours compromise an individual's ability to relate to others authentically[52] and can compromise the quality of relationships.

Organisational Impact

The impact of employee discrimination and exclusion is not limited to the person on the receiving end of such behaviours. In this section we explore the organisational costs of prejudiced behaviours.

Performance Impacts

The extensive negative impact of discrimination and exclusion on an individual has a significant effect on their performance which in turn affects the performance of their team and the organisation as a whole. When individuals experience the cognitive drain of concealment or the social withdrawal that protects them from discrimination, they are more likely to engage in physical withdrawal such as absenteeism.[53]

Turnover Impacts

Unsurprisingly individuals report an increased likelihood to quit in response to perceived discrimination at work.[54] The impact of this is twofold. First, the process of recruiting an employee to replace a leaver is costly. In the general context of conflict at work, ACAS estimates UK recruitment costs at £2.6bn per year and the cost of lost output as new employees get up to speed at £12.2bn per year.[55]

The second impact of an increased likelihood for victims of discrimination to resign relates to organisations' competitive advantage. Losing talent because of people's experiences of discrimination has an immediate impact on creativity and innovation. Over the longer term an outflow of talent can contribute to a growing reputation for being an unfair employer. Research has indicated that 70% of LGBTQ+ job seekers consider a company's reputation among the LGBTQ+ community when seeking or accepting employment[56] and therefore a bad reputation for discrimination or leaving negative behaviours unchallenged may hinder an organisation's competitive advantage.

Culture of Silence

Organisations tend to say very little about LGBTQ+ outside of Pride Month. Leaders can feel uncomfortable discussing the topic or maybe even think that it is no longer an issue in their organisation.

However, organisational silence reduces the likelihood that an individual will perceive it safe to disclose their identity, to report discriminatory behaviour and to be their true selves in social interactions with colleagues. This further limits individual and organisational efforts to learn and act.

Key Points

There is a considerable impact on LGBTQ+ individuals and community of the stigmatisation, stereotyping and discrimination that they experience.

At an individual level, it has a considerable impact on career choices and aspirations. Individuals are more likely to choose careers where they think that they will be accepted for who they are this also means closing the doors on other occupations where they may have greater abilities and interest. Members of the LGBTQ+ community are also more likely to be unemployed.

Identity management is something that members of the LGBTQ+ Community have to consider carefully. In particular individuals will be careful about who they share their identity with.

There are consequences for making known one's sexual identity to others in the organisation and the research shows that it can lead to greater discrimination.

The process of trying to keep one's identity hidden, however, has an impact on the individuals themselves. The energy that it takes to monitor the environment and to adapt one's behaviour is considerable. This has an impact in terms of people's image and identity, relationships with other people and performance on the job. Ultimately the process of being stereotyped and being discriminated against has an impact in terms of people's well-being, both physical and psychological.

We also need to remember that there is an organisational impact also. Lack of engagement at work means that people are more likely to leave, and the organisation not only loses an employee but picks up the tab for finding the replacement.

Notes

1 Schraepen, T. (2022). Do LGBTQIA+ People Face EU Labour Market Discrimination? *Bruegel Blog*, 26 September.
2 Government Equalities Office: National LGBT Survey. https://assets.publishing. service.gov.uk/media/5b3cb6b6ed915d39fd5f14df/GEO-LGBT-Survey-Report. pdf
3 Grant, J.M., Mottet, L.A., Tanis, J., Harrison, J., Herman, J.L., & Keisling, M. (2011). *Injustice at Every Turn: A Report of the National Transgender Discrimination Survey*. Washington: National Center for Transgender Equality and National Gay and Lesbian Task Force.
4 Schraepen, T. (2022). Do LGBTQIA+ People Face EU Labour Market Discrimination? *Bruegel Blog*, 26 September.
5 Grant, J.M., Mottet, L.A., Tanis, J., Harrison, J., Herman, J.L., & Keisling, M. *Injustice at Every Turn: A Report of the National Transgender Discrimination Survey*. Washington: National Center for Transgender Equality and National Gay and Lesbian Task Force, 2011.
6 Bachmann, C.L., & Gooch, B. (2018). *LGBT in Britain: Work Report*. Stonewall.
7 Drydakis, N. (2021). Social Rejection, Family Acceptance, Economic Recession, and Physical and Mental Health of Sexual Minorities. *Sexuality Research and Social Policy*, pp. 1–23.
8 Winderman, K., Martin, C.E., & Smith, N.G. (2018). Career Indecision Among LGB College Students: The Role of Minority Stress, Perceived Social Support, and Community Affiliation. *Journal of Career Development*, 45(6), pp. 536–550. https://doi.org/10.1177/0894845317722860
9 Lehtonen, J. (2008). Career Choices of Lesbian Women. *Journal of Lesbian Studies*, 12, pp. 97–102.
10 Boatwright, K.J., Gilbert, M.S., Forrest, L., & Ketzenberger, K. (1996). Impact of Identity Development Upon Career Trajectory: Listening to the Voices of Women. *Journal of Vocational Behavior*, 48(2), pp. 210–228.
11 Ng, E.S., Schweitzer, L., & Lyons, S.T. (2012). Anticipated Discrimination and a Career Choice in Nonprofit A Study of Early Career Lesbian, Gay, Bisexual, Transgendered (LGBT) Job Seekers. *Review of Public Personnel Administration*, 32(4), pp. 332–352.

12 Koenig, A.M., Eagly, A.H., Mitchell, A., & Ristikari, T. (2011). Are Leader Stereotypes Masculine? A Meta-Analysis of Three Research Paradigms. *Psychological Bulletin*, 137(4), pp. 616–642. https://doi.org/10.1037/a0023557

13 Correll, S.J. (2004). Constraints into Preferences: Gender, Status, and Emerging Career Aspirations. *American Sociological Review*, 69, pp. 93–113.

14 Chambers, N., Kashefpakdel, E.T., Rehill, J., & Percy, C. (2018). Drawing the Future: Exploring the Career Aspirations of Primary School Children From Around the World. *International Journal of Physical Education*. www.educationandemployers.org/wp-content/uploads/2018/01/DrawingTheFuture.pdf

15 Lippa, R. (2010). Sex Differences in Personality Traits and Gender-Related Occupational Preferences Across 53 Nations: Testing Evolutionary and Social-Environmental Theories. *Archives of Sexual Behavior*, 39(3), pp. 619–636.

16 Ellis, L., Ratnasingam, M., & Wheeler, M. (2012). Gender, Sexual Orientation, and Occupational Interests: Evidence of Their Interrelatedness. *Personality and Individual Differences*, 53(1), pp. 64–69.

17 Lippa, R. (2008). Sex Differences and Sexual Orientation Differences in Personality: Findings from the BBC Internet Survey. *Archives of Sexual Behavior*, 37(1), pp. 173–187.

18 Blashill, A.J., & Powlishta, K.K. (2009). Gay Stereotypes: The Use of Sexual Orientation as a Cue for Gender-Related Attributes. *Sex Roles*, 61, pp. 783–793.

19 Schneider, M.S., & Dimito, A. (2010). Factors Influencing the Career and Academic Choices of Lesbian, Gay, Bisexual, and Transgender People. *Journal of Homosexuality*, 57(10), pp. 1355–1369. https://doi.org/10.1080/00918369.2010.517080

20 Willis, P. (2010). Connecting, Supporting, Colliding: The Work-Based Interactions of Young LGBQ-Identifying Workers and Older Queer Colleagues. *Journal of LGBT Youth*, 7, pp. 224–246.

21 Bachmann, C.L., & Gooch, B. (2018). *LGBT in Britain: Work Report*. Stonewall.

22 HRC Report. 2009. https://assets2.hrc.org/files/assets/resources/DegreesOfEquality_2009.pdf

23 Köllen, T. (2013). Bisexuality and Diversity Management—Addressing the B in LGBT as a Relevant 'Sexual Orientation' in the Workplace. *Journal of Bisexuality*, 13(1), pp. 122–137. https://doi.org/10.1080/15299716.2013.755728

24 Mennicke, A., Gromer, J., Oehme, K., & MacConnie, L. (2016). Workplace Experiences of Gay and Lesbian Criminal Justice Officers in the United States: A Qualitative Investigation of Officers Attending a LGBT Law Enforcement Conference. *Policing and Society*, 0(0), pp. 1–18. https://doi.org/10.1080/10439463.2016.1238918

25 Colgan, F., Creegan, C., Mckearney, A., & Wright, T. (2008). Lesbian Workers: Personal Strategies Amid Changing Organisational Responses to 'Sexual Minorities' in UK Workplaces. *Journal of Lesbian Studies*, 12(1), pp. 31–45. https://doi.org/10.1300/10894160802174284

26 Ragins, B.R. (2008). Disclosure Disconnects: Antecedents and Consequences of Disclosing Invisible Stigmas Across Life Domains. *Academy of Management Review*, 33, pp. 194–215.

27 Hewlett, S.A., Sears, T., Sumberg, K., & Fargnoli, C. (2013). *The Power of "Out" 2.0: LGBT in the Workplace*. Center for Talent Innovation.

28 Malterud, K., & Bjorkman, M. (2016). The Invisible Work of Closeting: A Qualitative Study about Strategies Used by Lesbian and Gay Persons to Conceal Their Sexual Orientation. *Journal of Homosexuality*, 63(10), pp. 1339–1354. https://doi.org/10.1080/00918369.2016.1157995

29 Rood, B.A., Maroney, M.R., Puckett, J.A., Berman, A.K., Reisner, S.L., Pantalone, D.W. (2017). Identity Concealment in Transgender Adults: A Qualitative

Assessment of Minority Stress and Gender Affirmation. *American Journal of Orthopsychiatry*, 87(6), pp. 704–713. https://doi.org/10.1037/ort0000303

30 Omurov, N. (2017). Identity Disclosure as a Securityscape for LGBT People. *Psychology in Russia: State of the Art*, 10(2), pp. 63–86.

31 Carragher, D.J., & Rivers, I. (2002). Trying to Hide: A Cross-National Study of Growing Up for Non- Identified Gay and Bisexual Male Youth. *Clinical Child Psychology and Psychiatry*, 7(3), pp. 457–474.

32 Popova, M. (2018). Inactionable/Unspeakable: Bisexuality in the Workplace. *Journal of Bisexuality*, 18(1), pp. 54–66. https://doi.org/10.1080/15299716.2017. 1383334

33 Wegner, D.M., Schneider, D.J., Carter, S.R., & White, T.L. (1987). Paradoxical Effects of thought Suppression. *Journal of Personality and Social Psychology*, 53(1), pp. 5–13. https://doi.org/10.1037/0022-3514.53.1.5

34 Major, B., & Gramzow, R.H. (1999). Abortion as Stigma: Cognitive and Emotional Implications of Concealment. *Journal of Personality and Social Psychology*, 77(4), pp. 735–745. https://doi.org/10.1037/0022-3514.77.4.735

35 Uysal, A., Lin, H.L., & Knee, C.R. (2010). The Role of Need Satisfaction in Self-Concealment and Well-Being. *Personality and Social Psychology Bulletin*, 36(2), pp. 187–199. https://doi.org/10.1177/0146167209354518

36 Kerns, J.G., Cohen, J.D., MacDonald, A.W. 3rd, Cho, R.Y., Stenger, V.A., Carter, C.S. (2004, February 13). Anterior Cingulate Conflict Monitoring and Adjustments in Control. *Science*, 303(5660), pp. 1023–1026. https://doi.org/10.1126/science.1089910

37 Cortopassi, A.C., Starks, T.J., Parsons, J.T., & Wells, B.E. (2017). Self-Concealment, Ego Depletion, and Drug Dependence Among Young Sexual Minority Men Who Use Substances. *Psychology of Sexual Orientation and Gender Diversity*, 4(3), pp. 272–281. https://doi.org/10.1037/sgd0000230

38 Nouvilas-Pallejà, E., Silván-Ferrero, P., Fuster-Ruiz de Apodaca, M.J., & Molero, F. (2018). Stigma Consciousness and Subjective Well-Being in Lesbians and Gays. *Journal of Happiness Studies: An Interdisciplinary Forum on Subjective Well-Being*, 19(4), pp. 1115–1133. https://doi.org/10.1007/s10902-017-9862-1

39 Mereish, E.H., & Poteat, V.P. (2015). A Relational Model of Sexual Minority Mental and Physical Health: The Negative Effects of Shame on Relationships, Loneliness, and Health. *Journal of Counseling Psychology*, 62(3), pp. 425–437. https://doi.org/10.1037/cou0000088

40 Quinn, D.M., & Chaudoir, S.R. (2015). Living with a Concealable Stigmatized Identity: The Impact of Anticipated Stigma, Centrality, Salience, and Cultural Stigma on Psychological Distress and Health. *Stigma and Health*, 1(S), pp. 35–59. https://doi.org/10.1037/2376-6972.1.S.35

41 McAdams-Mahmoud, A., Stephenson, R., Rentsch, C., Cooper, H., Arriola, K.J., Jobson, G., McIntyre, J. (2014). Minority Stress in the Lives of Men Who Have Sex with Men in Cape Town, South Africa. *Journal of Homosexuality*, 61(6), pp. 847–867. https://doi.org/10.1080/00918369.2014.870454

42 Hendy, H.M., Joseph, L.J., & Can, S.H. (2016). Repressed Anger Mediates Associations between Sexual Minority Stressors and Negative Psychological Outcomes in Gay Men and Lesbian Women. *Journal of Gay & Lesbian Mental Health*, 20(3), pp. 280–296. https://doi.org/10.1080/19359705.2016.1166470

43 Newheiser, A.-K., Barreto, M., & Tiemersma, J. (2017). People Like Me Don't Belong Here: Identity Concealment is Associated with Negative Workplace Experiences. *Journal of Social Issues*, 73(2), pp. 341–358. https://doi.org/10.1111/josi.12220

44 Ullrich, P.M., Lutgendorf, S.K., & Stapleton, J.T. (2003). Concealment of Homosexual Identity, Social Support and CD4 Cell Count Among HIV-Seropositive

Gay Men. *Journal of Psychosomatic Research*, 54(3), pp. 205–212. https://doi.org/10.1016/S0022-3999(02)00481-6

45 Beals, K.P., Peplau, L.A., & Gable, S.L. (2009). Stigma Management and Well-Being: The Role of Perceived Social Support, Emotional Processing, and Suppression. *Personality and Social Psychology Bulletin*, 35(7), pp. 867–879.

46 Smyth, J.M., Stone, A.A., Hurewitz, A., & Kaell, A. (1999). Effects of Writing about Stressful Experiences on Symptom Reduction in Patients with Asthma or Rheumatoid Arthritis: A Randomized Trial. *JAMA: Journal of the American Medical Association*, 281(14), pp. 1304–1309. https://doi.org/10.1001/jama.281.14.1304

47 Cole, S.W., Kemeny, M.E., Taylor, S.E., & Visscher, B.R. (1996). Elevated Physical Health Risk Among Gay Men Who Conceal Their Homosexual Identity. *Health Psychology*, 15(4), pp. 243–251. https://doi.org/10.1037/0278-6133.15.4.243

48 Legate, N., Ryan, R.M., & Weinstein, N. (2012). Is Coming Out Always a "Good Thing"? Exploring the Relations of Autonomy Support, Outness, and Wellness for Lesbian, Gay, and Bisexual Individuals. *Social Psychological and Personality Science*, 3(2), pp. 145–152. https://doi.org/10.1177/1948550611411929

49 Pennebaker, J.W. (ed.). (1995). *Emotion, Disclosure, & Health*. American Psychological Association. https://doi.org/10.1037/10182-000

50 Crocker, J., & Major, B. (1989). Social Stigma and Self-Esteem: The Self-Protective Properties of Stigma. *Psychological Review*, 96(4), pp. 608–630. https://doi.org/10.1037/0033-295X.96.4.608

51 Frable, D.E., Platt, L., & Hoey, S. (1998, April) Concealable Stigmas and Positive Self-Perceptions: Feeling Better around Similar Others. *Journal of Personality and Social Psychology*, 74(4), pp. 909–922. https://doi.org/10.1037//0022-3514.74.4.909

52 Critcher, C.R., & Ferguson, M.J. (2014). The Cost of Keeping It Hidden: Decomposing Concealment Reveals What Makes It Depleting. *Journal of Experimental Psychology: General*, 143(2), pp. 721–735.

53 Volpone, S.D., & Avery, D.R. (2013). It's Self Defense: How Perceived Discrimination Promotes Employee Withdrawal. *Journal of Occupational Health Psychology*, 18(4), pp. 430–448. https://doi.org/10.1037/a0034016

54 Volpone, S.D., & Avery, D.R. (2013). It's Self Defense: How Perceived Discrimination Promotes Employee Withdrawal. *Journal of Occupational Health Psychology*, 18(4), pp. 430–448. https://doi.org/10.1037/a0034016

55 www.acas.org.uk/costs-of-conflict

56 Hewlett, S.A., Sears, T., Sumberg, K., & Fargnoli, C. (2013). *The Power of "Out" 2.0: LGBT in the Workplace*. Center for Talent Innovation.

Chapter 7

The Venn Diagrams of LGBTQ+ Intersectionality

Disability, Race and Age

Introduction to Intersectionality

> I love Venn diagrams. I love Venn diagrams. You know-the three circles? I love Venn diagrams . . . I asked my team from which states are we seeing attacks on reproductive health care, attacks on voting rights, and attacks on LGBTQ+ rights. Texas, Florida, Georgia okay so there you go. So, you are not surprised to know but there is an overlap.[1]

In sharing the love of van diagrams, Kamala Harris, the vice president of the United States from 2016 to 2020, was also describing intersectionality—it's the area in the centre of the diagram which shows where the commonalities are on a number of topics.

African-American women faced challenges in having their discrimination cases recognized because the available data showed no evidence of discrimination against women as a group or African-American individuals as a group. However, the reality was that white women and African-American men were being treated fairly, while African-American women experienced specific discrimination based on the intersection of race and gender.[2] In response, legal scholar Kimberlé Crenshaw introduced the concept of intersectionality to highlight this issue. Since then, the concept has sparked significant discussion and deeper exploration.

Autobiographies of LGBTQ+ individuals have become a notable space for discussing intersectionality serving as helpful case studies which show the subtle effects of one identity on another.

However, it is not always used in a useful way. Within organisations intersectionality is in danger of becoming a new buzzword, making speakers appear up to date, even ahead of the game. It is also sometimes used as a weapon: "As a working class, LGBTQ+ person of colour, you can't possibly understand how I feel." We have also seen it used as a deflection strategy—to avoid talking about one thing in order to move the discussion onto something that the individual is more comfortable with and concerned about. It hasn't yet translated adequately into organisational policy or even into the way that they analyse the data.

DOI: 10.4324/9781003489580-9

In this chapter we want to explore what intersectionality means and what it can bring to our understanding of LGBTQ+ experiences at work and in society.

A recent literature review[3] identified three major approaches to examining issues of discrimination, diversity and inclusion:

- The singular approach involves looking at each topic in isolation. For example, LGBTQ+ and disability will be looked at separately and so implying that they have nothing in common.
- The additive approach is where research is carried out separately for different characteristics but then attempts will be made to connect them in some way. Generally speaking, the conclusion will be that the level of discrimination that someone with multiple minority identities experiences will be greater than those with one such identity. This is often referred to as double disadvantage.
- The intersectional approach involves looking at multiple identities and analysing how they interact with one another.

One particularly helpful theory when it comes to intersectionality is IPT, or Identity Process Theory.[4] The way we create and construct an identity for ourselves involves two major processes: first, assimilation—accommodation and second, evaluation.

Assimilation—accommodation refers to the ways in which we take on new information and insights about ourselves. For example, it could relate to the acknowledgement to oneself that "I am gay." This realisation, albeit at this stage in private, needs to be considered in the light of other identities, for example, disability, ethnicity and religion which can come into conflict with one another.

Accommodation occurs where these different identities are accepted and a way of making sense of them together is formed. This could involve incorporating them into one's identity and making them whole. It could also mean accepting that they will need to be seen as separate, leading to a break with one community or group.

Evaluation is the process of giving new meaning to the identities, for example, recognising one's sexual identity as positive and, in the light of that, potentially reassessing other identities.

According to IPT we are seeking to achieve a sense of:

- Continuity across time and environments
- Distinctiveness from other people, something which makes people feel unique in some way
- Self-efficacy, a sense of control and competence in the way they live their lives
- Self-esteem, recognising that one has some value and personal worth
- Belonging and inclusion

- Purpose and meaning in our lives
- Psychological coherence, so that different aspects of one's identity can be properly brought together as a whole.

The theory helps to understand the disturbance we feel when one important facet of our identity is either ignored or not accepted by those around us.

In this chapter we will be looking at LGBTQ+ and

- Disability
- Race
- Age

LGBTQ+ and Disability

Introduction

This section won't be looking at the models of disability, for example, the social construction model and the medical model, which provide a framework for the ways in which disability is viewed. This has been extensively written about and while the medical model and social construction model are the ones that are most widely known there are others that exist.[5]

Instead, our focus is the overlap between LGBTQ+ and disability.

Many countries have anti-discrimination legislation covering disability. What may not be so widely known is that the right to sexual well-being is also covered by legislation. Article 8 of the UK's Human Rights Act of 2008 protects people's right to respect your family and private life. The European Court of Human Rights has established that this includes recognition and protection of "sexual autonomy, confidentiality, dignity, forming and maintaining personal relationships and allowing them to develop."[6] Furthermore, the International Convention on the Rights of Persons with Disabilities (2006)[7] also recognises the importance of disabled people being able to engage fully and be included in their communities. The importance of sexual expression to sexual health was also recognised by the World Health Organisation in 2006.[8]

We mention this because it is important to recognise that sexual well-being and sexual expression are significant factors in our overall well-being. When it comes to people with disabilities therefore our beliefs about their sexual identity and expression represent obstacles to people being able to be their authentic selves, but we may also be failing to live up to the spirit as well as the letter of legislation and the various conventions.

Despite being two distinct identities, disability and LGBTQ+ share factors in common when it comes to perception.

Until relatively recently homosexuality was seen as some form of degenerative condition. Psychiatrists and psychoanalysts treated it as if it were some form of illness. For some people there still is the persistent belief that if this is an illness then there must be a cure.

Perhaps the most significant part of an identity is the naming of it. Insulting, derogatory and demeaning terms have been used to describe LGBTQ+ individuals and disabled people and both groups have had to struggle to have their own identity labels accepted.

Just as there is an assumption that everyone is heterosexual, famously described by Adrienne Rich as compulsory heterosexuality,[9] this same idea has been applied to disability and so provides the concept of compulsory ableism.[10] Without necessarily making it explicit, we expect and want people around us, including in our workplaces, who appear to be "normal." Members of the LGBTQ+ community and disabled people represent deviations from normality.

It has been estimated that 80% of disabilities are not visible.[11] Therefore many disabled people have to make a decision about whether they will conceal their disability or reveal it to others. This is very similar to the coming out process that LGBTQ+ people experience. For LGBTQ+ people with a disability, however, the process is even more difficult and complicated. Not only will they have a dual coming out process to deal with, but they may lack support from both the LGBTQ+ and the disability communities and subcultures.[12]

They are also negatively perceived. "Homosexuals and cripples have historically been called 'sinners', 'evil' or 'defective'."[13] Both identities feel a sense of isolation particularly when they are mistreated by their own families and the communities in which they live.

Research

Research into LGBTQ+ has rarely looked at the experiences of disabled people which render them invisible and leaves a gap in our knowledge and understanding which significantly impacts employment and access to services.

One of the first pieces of research to look at sexual orientation and disability was carried out by the Irish National Disability Authority in 2005.[14] It's testimony to that work, and to the lack of progress on this topic since, that many of the observations made in the report remain very relevant today.

The report identified the way that disability and sexuality interact with one another in a way which creates greater stress for LGB disabled people.

> They are individually, and collectively, subject to prejudices based on the "normalising" principles of a nondisabled, largely heterosexual mainstream population. The combination of this misconception with societal and cultural prejudices towards persons who are lesbian, gay or bisexual creates a population within society who face a double prejudice for acceptance and equal rights—as people with disabilities, and as lesbian, gay or bisexual individuals.[15]

It took a long time for researchers in the social sciences to look at disability and sexuality, because they were treated like two separate and unrelated topics. Change began to occur slowly in the 1970s when "the field of rehabilitation discovered sex."[16]

More research is now being conducted looking at the intersection between disability and sexuality. However, it seems that this is still relatively rare, and the more common approach is to ignore the topic of disability altogether.[17]

The research conducted by the National Disability Authority in Ireland, using as a basis data collected from the UK, observed that 10% of disabled people are lesbian or gay.

This would put the number of people of sexual minorities among disabled people higher than that of the general population. Other research has also reached the same conclusion. For example, in Australia 4.4% of disabled people identified as transgender, with 3% having a preference for terms other than "male" or "female" to describe their sexual identity or gender identity.[18]

The conclusion of some disability advocates that sexual minorities are more likely to be found among disabled people than in the general population has been questioned by some, but it does appear to be a consistent finding from different countries.

Another way of looking at the data, however, would lead to the conclusion that sexual minorities are more likely to have a disability. Analysis of the statistics in the United States found that "Disability as a whole in the LGBT community was 156% more prevalent than non-LGBT people, and this rises to 281% more prevalent for transgender people."[19]

In Australia, in research carried out looking at sexual orientation of men, over 90% (93.2%) of able-bodied men said that they were heterosexual. For men with a disability, however, the figure was lower at 87.7%. More than three times as many disabled people said that they were bisexual (3.9%) compared to people without a disability (1.2%). Other key areas of difference between able-bodied and disabled men were in terms of the number of disabled people who were unsure of their sexual identity and who they were attracted to.[20]

The other big difference was in the number of people who had a disability who stated that they had no sexual attraction (3.8% compared to 0.8%). The reasons for no sexual attraction may well be related, the research speculated, to the fact that there may have been fewer chances to discover their own responses to others and to develop a sense of who they are. Environmental factors can also play a part due to the presence of others around disabled more often, for example, carers, and so have less privacy.

Issues and Obstacles Faced by LGBTQ+ People with a Disability

The negative views of people with disabilities are nothing new. The reasons for this have been related to two specific forms of anxiety that people without a disability have: existential anxiety and aesthetic anxiety. There are also powerful negative stereotypes about disabled people which we also discuss.

Existential Anxiety

In ancient Greece and Rome, disability was seen as a punishment from the gods and disabled children would be abandoned and left to die in the streets. In the 19th and 20th centuries, there was legislation, known as the "ugly laws," to prevent disabled people from being seen in public.[21] Attitudes such as these led to the involuntary sterilisation of people with disabilities, particularly those with learning disabilities, a practice that was routinely carried in many countries until the mid-20th century.[22]

One of the reasons for these brutal policies is referred to as existential anxiety.

Meeting a disabled person reminds us how close we are to physical infirmity or even death. It confronts us with the fact that the barrier to having some form of impairment may literally be only skin deep.[23] If our skin is punctured in the wrong place or by the wrong implement, we could find ourselves with an illness leading to some form of long-term condition. These uncomfortable, even distressing, thoughts can be prompted by the presence of a disabled person. Better therefore to have these people more distant from us so we don't have to be confronted with our own mortality.

Aesthetic Anxiety

Another form of anxiety related disabled people is called aesthetic anxiety. Here the disabled people, in particular those with some form of physical impairment, will be shunned because they don't meet the narrowly defined idea of physical attractiveness. Disabled men will be seen as weak and vulnerable; disabled women are not seen as meeting the feminine ideal. Furthermore in the media disabled people are represented as some form of evil, strangeness and a distortion of what is expected in the civilised society.[24]

Identity

> When meeting someone in person, my disability shows itself, my wheelchair introduces itself, even before I can introduce myself. With a long struggle under my belt, my lesbianism is now more readily visible.[25]

This quote from a participant in research shows how having a clearly identifiable disability captures our attention so powerfully that it leaves us with no room to consider other aspects of a person's identity. Known as the spread phenomenon,[26] this is part of the reason why awkwardness and tension can exist in the interactions between disabled people and those who are not. It also means that other parts of the person's identity are not considered including their sexuality.

Asexuality

The National Disability Authority of Ireland's ground-breaking 2005 report referred to the "desexualisation" (p12) of disabled people, that is, that they are thought of as having no interest in sex. Sexual orientation therefore is of little significance, but when it is considered disabled people tend to be thought of as straight.

Assumptions are also made that impairments will lead to sexual dysfunction. For those with intellectual disabilities or psychiatric conditions, concern will be expressed that they lack the social awareness to be able to make appropriate decisions about the suitability of a relationship partner.[27]

Furthermore, those providing services to disabled people did not provide ways for their clients to express themselves sexually. Underlying their actions were negative attitudes and prejudices towards disabled people and having expectations of the way that they should be conducting themselves.[28] This applies particularly to those with a learning disability and who also have a gender identity disorder.[29]

Some physicians find it hard to imagine how somebody with a severe disability, for example, spinal cord injury (SCI) could be in a relationship and so the advice given to the patient, directly or more subtly, would be that finding a sexual partner would be next to impossible and that any thoughts of having a family should be suppressed.[30]

However, research into those with SCI concludes that "a man or a woman who had positive and pleasurable sexual experiences prior to injury would approach sexual function after injury with a positive attitude if there is a partner present with whom he or she has a good relationship"[31]

Infantilism

People with disabilities can find themselves being patronised and treated like a child, becoming infantilised, where a person "is treated in a paternalistic, compromising and ultimately disempowering way."[32] As we do with children, the topic of sex is avoided altogether.

Impact

Abuse and Violence

One of the effects of these attitudes and stereotypes is that disabled people are subject to greater stigmatisation, discrimination and marginalisation than those without a disability. They are also more likely to be subject to abuse and violence. The situation becomes even more difficult and dangerous for disabled people with a sexual minority identity—an observation that is seen to be true in different countries around the world.

A study in England found that crimes against disabled people are under-reported and that perpetrators are more likely to be someone from their own family or a carer.[33] Studies in India have found high rates of violence against women with disabilities, lesbians and female sex workers.[34]

In Australia, LGBT people with a disability were more likely to have experienced some form of harassment or violence in the 12 months prior to completing the survey compared to those LGBT people without a disability (46% against 33%). The type of abuse experienced included verbal (32% versus 14%), written (e.g. emails or graffiti) (11% versus 5%), harassment (21% versus 14%), threats (13% versus 8%); sexual assault figures for lesbian and bisexual women without a disability were 2% and 5% for lesbian and bisexual women with a disability. It's important to appreciate that the levels of violence against sexual minorities are very high and that it is higher still for those who have a disability.[35]

A European Union-sponsored report reached the same conclusion. Homophobia and transphobia are worst in countries which criminalise same-sex relationships. And then within that, experiences of LGBTQ+ people with a disability are even worse.[36]

The situation becomes more extreme when trans- and gender-nonconforming people are considered. Trans- and gender-non-conforming people experienced discrimination related to gender identity when accessing relevant services. Once again with this identity group, experiences were worse for those people with a disability.[37]

Well-being

The damaging and widely held belief that disabled people are asexual has an impact on their well-being. It will not be understood by those around them that a physical impairment that an individual has does not mean that they also experience a loss of sexual desire. This is true of disabled people generally, but the feelings are heightened, and the impact much worse when we are looking at LGBTQ+ people with disabilities.[38,39] For those with gender identity disorder and learning disability, the impact on their well-being is greater again.[40]

Being able to express oneself is only one element of this, however. Having an emotional connection with someone else also has a significant impact on people's well-being.

Age also plays a significant part. A very large-scale survey in the United States with nearly 100,000 participants 96,992 found that older LGB people have a higher risk of disability, poor mental health, smoking and of excessive drinking than older heterosexuals.[41]

Isolation

One of the key themes in research from different is that disabled people who are also LGBTQ+ experience discrimination, rejection and exclusion from both communities as well as from mainstream society. The attitudes of the people closest to them had the biggest impact on disabled people being able to express their sexual feelings and identities, irrespective of their orientation. However, the impact is more negative for those people with a disability who are also from a sexual minority and more negative again for those with learning disabilities.[42]

This was identified by the National Disability Authority of Ireland in 2005. There was homophobia within the disability community, but there was also discrimination against disabled people by LGB groups. The emphasis on body culture also had an impact on the perceptions of people with disabilities.

For those people with physical disabilities a practical issue was the accessibility of the buildings in which people gathered as part of the social support networks. This remains true today with pride events in cities such as Amsterdam. A report into Canal Pride found that whist attitudes were more inclusive within the LGBTQ+ community the physical environment remained a barrier for some people with disabilities. Being unable to access social gatherings like this, due to the physical constraints or the attitudes of others, leads disabled people who are also of a sexual minority to feel even more excluded.[43]

Conclusion

The overall conclusion that we take from our overview of the research on LGBT and disability is that this is an area that has been ignored in the research literature and in organisational practice. Because of the long-standing if not ancient attitudes towards disability, those who are disabled are subject to high levels of discrimination, stigmatisation and stereotyping. Those disabled people who also are from a sexual minority experience even more exclusion and discrimination than those who are heterosexual.

Returning to the identity process theory, it becomes very difficult for disabled people from the sexual minority to be able to incorporate all those elements into a coherent whole if those around them refuse to accept the legitimacy of all aspects of their identity. This includes not just friends, family and society generally but also the LGBTQ+ community itself. The sense of isolation that disabled LGBTQ+ people experience has an enormous impact on their health and well-being.

It is clear from this analysis that organisations need to ensure that disability is incorporated into the LGBTQ+ policies and that LGBTQ+ is incorporated within their disability policies. Without doing this we will be continuing the singular approach and in doing so we will fail to identify the extreme difficulties that this minority within a minority faces.

LGBTQ+ and Race

Introduction

"If black women were free, it would mean that everyone else would have to be free since our freedom would necessitate the destruction of all of the systems of oppression"—so read the statement issued by the Combahee River Collective (CRC).[44] The CRC was founded in 1977 by Barbra Smith, Beverley Smith and Demita Frazier, who, as black lesbians, found themselves in a position where they were either fighting against the oppression faced by black people or by women. In effect, however, they had to make a choice between having to challenge racism faced by black men or the misogyny and sexism faced by white women. Their own particular identities and experiences were effectively ruled unimportant by these two movements. By establishing the CRC, they highlighted the ways in which black women had been ignored.

The CRC was Marxist and extended the analysis of capitalism to include the oppression of black women. They began running consciousness-raising groups and shortly afterwards issued the Combahee River Collective Statement. The creation of the CRC proved to be an influential moment, leading to significant changes in the way that we view equality in workplaces by adopting an intersectional approach.

Data on Population Size

A 2012 Gallup survey in the United States found that of the 3.5% who identified as LGBT, a third (33%) were racial minorities. In 2016 Gallup found that the figure was 4.1% (approximately 10 million adults), and of these, minorities now represented 40% (4 million adults). The numbers of people identifying as white and LGBT increased from 3.2% to 3.6%, African Americans increased from 4.4% to 4.6%, Hispanics from 4.3% to 5.4%, and Asians showed the biggest increase from 3.5% to 4.9%.

These figures show a shift in social attitudes towards growing acceptance, as well as an increased visibility of different sexual and gender identities. This was a significant increase in such a short period and was largely attributed to the millennial generation, those born between 1981 and 1996, who increased from 5.8% to 7.3%. Millennials were twice as likely to identify as being LGBT as other generations.[45]

Treating the LGBTQ+ community as homogeneous therefore in terms of race does everyone a disservice. Data such as this was used initially to help service providers to understand the service users better and to improve their advice, guidance and support.

According to the 2021 census data for England and Wales, approximately 90% of the people (89.4%) were straight. Just over 3% of people were LGB or "other sexual orientation."

After the 2011 census, it was estimated that 400,000 people who identify as being LGBTQ+ are also members of an ethnic minority group (Stonewall). This is approximately 40% of the UK's 1.1 million LGB population, which corresponds to the calculation from the United States. In both countries therefore minorities form a greater proportion of the LGBTQ+ population than they do of the population as a whole.

Obstacles and Issues Faced by LGBTQ+ Ethnic Minorities

Identity Development and Community Rejection

Comparisons have been drawn between the development of a sexual identity with that of a racial or ethnic identity.

Sexual identity has been conceived of as being developed through a series of stages. This starts with an attraction towards people of the same sex and the realisation that this doesn't match with society's expectations. A period of confusion occurs where the individual tries to understand what this means for them as well as becoming aware of the dangers that are posed by coming out. There will be a stage of trying to conform and be someone different to who they actually are, for example, by dating or marrying someone of the opposite sex. In terms of Identity Process Theory, there are a number of discontinuities and dislocations which leave people feeling confused about who they actually are. Eventually, by coming into contact with members of the LGBTQ+ community, they can reach acceptance of their own sexual identity. They can see ways in which they are unique and can take some control of the way in which they live their life. In a model like this, the final stage is represented not only by the individual accepting who they are but also by having this recognised and acknowledged by others.[46]

Much of the work on identity development when it comes to sexuality applies to anyone irrespective of ethnicity. With minorities, there is also the development of a racial or ethnic identity, which is occurring in parallel. Children growing up start to become aware of the ethnic group to which they belong, and the attitudes others hold about them, positively or negatively. Their own ethnic identity will start to develop its own more distinct profile, which becomes clearer typically after events they experience in their lives. These events could be a single encounter or a series of smaller ones from which they discern a pattern relating to their ethnicity.

It is during adolescence that these two identities begin to take shape, but the processes are different. In terms of racial or ethnic identity the process is supported by members of their families, relations, friends and the community in general. The process of sexual identity is more private initially, but the process of recognition and acceptance relies upon having contact with LGBTQ+ individuals and organisations, as well as learning from peers and the internet.

Generally speaking, being involved in and accepted by one's ethnic community is beneficial to well-being. However, for ethnic minority LGBTQ+ individuals, this is particularly difficult. Ethnic minority LGBTQ+ youth are less likely to have come out to their parents compared to their white counterparts[47,48] and, on coming out, individuals may face higher levels of resentment and hostility[49,50] as well as greater sexual orientation hate crime than White people.[51,52]

Another source of external pressure on LGBTQ+ individuals from minority groups is the very explicit belief in the community that heterosexual relationships are considered "normal" and an expectation of marriage to someone of the opposite sex. Not fulfilling this expectation can be a source of anguish and disappointment but may also have wider familial and social ramifications. Families could feel that they have been shamed, but more widely than that, it could also be seen as a threat to the whole of the community, with pressure being brought to isolate and expel the errant individual.[53] Whereas 9% of LGBTQ+ people who are Christian have experienced discrimination and poor treatment because of their faith, this rises to 20% of LGBTQ+ people who are adherents of other faiths, including Islam, Sikhism and Judaism.[54]

Knowing the consequences means that individuals won't come out and will agree to a heterosexual marriage in order to demonstrate their "normality" and commitment to societal norms.[55]

The combination of expectations, loyalty to the family and religious beliefs can all lead to someone feeling as if they will never be seen as representative of the community, which leads to a greater sense of isolation. The sense of loneliness will increase because these same pressures will mean that straight members of minority communities may be less willing to be seen as allies.

The sense of shame and worthlessness experienced will make people withdraw from situations and make themselves as invisible as possible. Individuals from minority groups are less likely than white people to:

- know or have had contact with anybody from the LGBTQ+ Community (42% compared to 53% of white people)
- have friends or family who are gay or lesbian (25% compared to 40%)
- have acquaintances in their community (20% compared to 30%).[56]

In the short term, this may help people, but in the longer term, it will be detrimental to their health and well-being. It means that they are less likely to reach out for support and are more likely to engage in risky behaviour, such as drug or alcohol abuse.[57,58]

Racism from the LGBTQ+ Community

> Sometimes it feels like I'm an outsider in a niche group. . . . It feels like I'm not accepted anywhere. With People of Color, I'm gay. With gay people, I'm a Person of Color. I'm always different even when I'm around people who I'm supposed to be like. It's hard to find your people.[59]

This quote from David, a gay black man, highlights another major problem that can be experienced by minority LGBTQ+ individuals-rejection from the white LGBTQ+ community.

Some of the key points that emerge from the research include:

- Experiencing insensitivity and micro-incivilities in gay restaurants and bars[60]
- Not being listened to by white LGBTQ+ people who lacked awareness or were not in the multiple forms of discrimination[61]
- Being stereotyped by their ethnic group[62]
- Not getting support from witnesses to racist acts[63]
- Being overlooked by LGBTQ+ campaigning organisations, who, without perhaps, realising it, focus on white people more.[64] (However, because of the external pressures placed upon them, minorities may be more concerned about taking up LGBTQ+ activism.)
- Experiencing forms of what has been called sexual racism, which refers to "sexual preferences" where minorities are seen as being less attractive.[65]
- Older African-American gay and bisexual men were more likely to experience ageism than their white counterparts.[66]
- Beliefs about the role that minorities could play in a sexual relationship. Black people being seen as more masculine would be expected to adopt the more dominant role, Asians were expected to be more submissive, and because they were viewed as more reserved, it would lead to sex that was less passionate and exciting.[67]
- Racist terms being used to describe minorities. For example, Asian men, who were attracted to other Asian men were referred to as "sticky rice." For those who are not Asian, but who are attracted to Asians the term "rice queen" is used. On dating apps references to "no rice" and "no curry" indicated the racial preferences of the user.[68]

A majority of LGBTQ+ ethnic minorities in the UK have experienced verbal and physical abuse from the LGBTQ+ community and many have reported that witnesses to the attacks did not intervene to support them. This can lead to long-term consequences in terms of individuals' mental health, including some who still experienced post-traumatic stress disorder (PTSD).[69]

All these factors and experiences mean that having been rejected by their own ethnic community, they don't necessarily find themselves being accepted by the LGBTQ+ community either, which only heightens the sense of isolation and lack of support.

Deracialisation

We reach conclusions about someone's identity based on stereotypes. If someone is speaking in a posh or upper-class accent, the people hearing it are more likely to reach the conclusion that they are white.[70] When people are asked to identify someone as masculine, they do it more quickly for black people than for white people. This is because stereotypically, black men are seen as more masculine than white or Asian men.[71] This shows that there is an interaction between different identities, which leads to different assumptions being formed and different biases.

In Western countries, when people are asked to consider a typical person of that nation, they will usually describe someone who is white, male and heterosexual, this being an expression of ethnocentrism, androcentrism and heterocentrism. When non-white groups are considered, the prototypical individual is still seen as being male and heterosexual.

Psychologists Christopher Petsko and Galen Bodenhausen were interested to find out what happens when people are asked to consider the characteristics of a minority person who is also gay. Because the assumption was that the prototypical person representative of that minority group is straight, when they were asked to categorise somebody who was both from a minority group and gay, they saw them as being more like the majority (i.e. white), and less like people from their ethnic group. In other words, being gay had the effect of "deracializing" them.[72]

The reasons for the de-racialisation differ for each ethnic group, so this isn't a universal process. For Hispanic people being gay creates tension with the stereotypes of that group, being religious and family oriented. For Asians, it is because the stereotype of them is of being traditional and somewhat reserved. The stereotype of gay men clashed with the hyper-masculine stereotype of black men. Not being seen as typical of their ethnic group then led to them being viewed as more white. This also had other consequences. As gay African Americans and Hispanics were seen as more white, are also viewed as more affluent than straight African Americans and Hispanics. (This effect did not occur for Asian people who are already stereotyped as being wealthier in any case). Less deracialisation occurs for white people, and this could be because most of the participants were white, and consequently they could see more variation and diversity in their own ethnic group.

Having more than one stigmatised identity doesn't necessarily lead to a doubling of the bias and discrimination that people experience. The interaction between different identities is more complex and subtle than we might think. The same deracialisation effect has been found in Sweden, where gay Arabs were treated more positively than heterosexual Arabs.[73]

As being American is associated with being white,[74] minority gay individuals are more likely to be viewed as a member of the ingroup and American.

In effect what people seem to be saying is "you're not like the others—you're more like one of us." This certainly seems to be the case when it comes to selection decisions, where minority gay applicants for jobs were discriminated against less than minority heterosexual applicants.[75]

On the other hand, if they are perceived as being more white, they may be viewed as not belonging to the minority in-group. They may receive less support and sympathy from members of their own ethnic group because they are seen as more like part of the majority.[76]

Impact

Well-being

Minorities in communities and families which demonstrate homophobia are more likely to internalise these attitudes and see themselves as unworthy and sinful. The emotions experienced will be guilt, shame and self-disgust.

The discrimination and isolation experienced by minority LGB individuals, compared to white LGB counterparts had the following impact:

- Greater drug use
- Less likely to be out
- Greater internalised homophobia
- Greater risk of suicide

Being accepted by those close to them, however, results in much more positive outcomes. People are more likely to be out and will receive support from others around them, which leads to more positive self-evaluation.[77]

Families, as you might expect, exert a considerable influence, and when family members are accepting of their son, daughter, sibling et cetera, it has been found that individuals are less likely to show depressive symptoms, have better general health and feel more supported in their lives.[78]

Pay

Much attention has rightly been given to the gender pay gap and there is increasing attention to the ethnicity pay gap. Little attention has been paid to the impact of sexual orientation, and even less to the intersections between gender, ethnicity and sexual orientation.

What we do know is that results are different for each subgroup for the LGBTQ+ community. Typically, gay men experience a wage penalty when it comes to pay, but lesbians potentially experience a wage premium,[79] with the same pattern being found for different ethnic groups.

In a fascinating and beautifully presented piece of research Del Rio and Alonso-Villar show the interactions between different characteristics.[80]

Black men, whether straight or gay, experienced a significant negative pay gap when compared to white heterosexual men. For black gay men, it was 25.8% less and for black heterosexual men the gap was slightly larger at 27.4%. Black gay men earn slightly more (1.5%) than black heterosexual men, but a lot less (13.5%) than white gay men.

Hispanic gay men earned less than all others in the comparisons: 3.9% less than Hispanic heterosexual men; 9.3% less than white, gay men; 21.7% less than white heterosexual men.

Hispanic, heterosexual men earned 17.7% less than white heterosexual men.

Asian gay men, as with Hispanic gay men, were paid less in all of the comparisons. They earned 3.1% less than Asian, heterosexual men, 1.5% less than white gay men and 13.8% less than white heterosexual men. Asian straight men earned 10.7% less than white straight men.

White lesbian women earned 8.2% more than white Heterosexual women, but 13.9% less than white gay men. Black lesbian women earned 1.2% more than black Heterosexual women, but 14% less than white lesbian women.

When compared to White, Heterosexual women, Hispanic and Asian lesbian, women, both and more, 5.4% for the former and 15.7 for the latter. Hispanic lesbian women earned 13.7% more than Hispanic, Heterosexual women, but 2.8% less than white lesbian women.

Asian lesbian women earned 7.5% more than white lesbian women and 12.8% more than Asian Heterosexual women.

There is, in other words, a complex interaction between sex, race and sexual orientation which doesn't necessarily always go in the direction of increased disadvantage occurring with multiple minority identities.

Conclusion

There is a complex interaction in terms of identity development that occurs in adolescence. Becoming aware of one's ethnicity is a public and shared experience. While this is occurring, individuals become aware of the negative views that are held of people who are not heterosexual.

An important difference between the experience of minorities and those in the majority is that LGBTQ+ individuals are not only seen as bringing shame on their families but are considered to be a threat to the community as a whole. As a consequence, as individuals become aware of their sexual identity, they will also feel that in terms of achieving some form of coherence, they may have to leave that community behind. There are significant consequences in terms of health and well-being in doing this. However, it would also appear that the LGBTQ+ Community has not always been as inclusive as it could've been leading to minorities feeling even more isolated with negative impacts on the health and well-being.

In terms of organisations, however, we need to be careful to just assume that having multiple minority identities leads to even greater discrimination against them.

In fact, what seems to occur is a deracialisation of individuals so that they are seen as more like members of the majority in group. It is important in organisations that race and LGBTQ+ are not seen as somehow separate from one another. What this analysis of the literature reveals is that they are totally bound up with one another.

LGBTQ+ and Older People

Introduction

Human development has typically been seen as a series of stages that someone goes through. Shakespeare famously had his seven ages of man which describes the behaviours associated with each age. This approach is increasingly being replaced by lifespan theories, which see ageing as a more fluid and creative process where people are making active decisions about how they will live and work throughout their lives.[81]

Some of the important ways in which people change and adapt as they get older include:

- Finding ways to compensate for the relative loss of certain capabilities, for example by using their experience to anticipate events and to solve problems;
- Placing greater emphasis on the emotional rewards that can be obtained from work, finding meaning in work and also passing on skills to a younger generation by coaching and mentoring them;
- Being prepared to create an environment which provides them with greater satisfaction, as opposed to constantly trying to adapt to fit their workplace;
- Being flexible in the way that work-related and career-related goals are achieved.[82]

Such approaches are having an impact in the workplace in terms of the way older generations are perceived. When it comes to consideration of sexual orientation, however, there are marked differences between the generations.

Research

LGBTQ+ by Generation

There are clear differences in the size of the LGBTQ+ population by generation. Researchers identify generations typically by using the following classification:

The Silent Generation born 1945 or earlier
Baby Boomers born between 1946 and 1964

Generation X born between 1965 and 1980
Millennials born between 1981 and 1996
Generation Z born between 1997 and 2012
   (polling companies vary in terms of exact date ranges but only slightly)

The 2024 Ipsos global survey of 26 countries found that 17% Generation Z identified as LGBT+, with the percentages declining with each subsequent age group: 11% of Millennials, 6% of Generation X and 5% of Baby Boomers.[83]

Data from individual countries reflects that overall pattern. The 2021 census in the UK,[84] using different age categories, found that the younger age groups were more likely to identify as LGB with numbers declining consistently for each of the older age groups:

16 to 24 years of age 6.91%
25 to 34 years of age 5.63%
35 to 44 years of age 3.5%
45 to 54 years of age 2.39%
55 to 64 years of age 1.59%
65 to 74 years of age 0.84%
75 years of age and above 0.37%

Lesbians and gay men were the majority within each age category apart from the 16- to 24-year-olds where people who were bisexual were the largest group.

In the United States, Gallup have been collecting data on sexual orientation since 2012.[85] The number of people identifying as LGBTQ+ has been increasing year on year, and in 2024 the number was 7.1%, more than double the figure for 2012 (3.5%). Once again there are considerable generational differences for those identifying as LGBT:

Generation Z 20.8%
Millennials 10.5%
Generation X 4 .2%
Baby Boomers 2.6%
Silent Generation 0.8% (Gallup refer to this group as "Traditionalists")

Since 2012 there has been little change in the numbers for the older age groups of the Silent Generation, Baby Boomers and Generation X. There has been an increase among Millennials from 5.8% in 2012 to 10.5% in 2024. The first members of Generation Z to reach adulthood was in 2017 when 10.5% identified as LGBT+. By 2024 this figure has sharply increased to 20.8%.

According to the Gallup data more than half of the people (57%) of those identifying as LGBT are bisexual.

Life Experiences

It's important to recognise that the experiences of LGBTQ+ individuals of older generations (Silent Generation, Baby Boomers and Generation X) differ markedly not only from heterosexual peers but also from the younger generations of the LGBTQ+ community.

Some of the key dates which it is important to remind ourselves of include:

- 1952—homosexuality is identified as a sociopathic personality disturbance in DSM-I
- 1968—in DSM-II it is still recognised as a mental disorder
- 1990—the World Health Organisation declassifies homosexuality as a mental disorder
- 2013—in DSM-5 it is removed as a mental disorder

This list doesn't contain any of the legislation which makes homosexual activity unlawful. For example, in the UK Section 28 of the Local Government Act of 1986 stated that it would be unlawful for any local authority to promote homosexuality in any way. This had an immediate impact on many organisations working with the LGBTQ+ community who found themselves without local government support or funding. Such actions would continue to make LGBTQ+ individuals feel that they were not "normal" and represented a threat to the rest of society by persuading straight people to forgo heterosexuality.

In the 1980s the HIV/AIDS epidemic meant that the stigmatisation of the LGBT community became significantly worse as was the sense of isolation. Individuals had to deal with friends who had been diagnosed with HIV/AIDS, and yet were unable to discuss their feelings at work for fear of being discriminated against.

People between 60 years of age and above today have lived through a period of intense suspicion, stigmatisation and discrimination. The levels of secrecy that people in these older age groups would have endured during these times have left their mark on them today.

Studies looking at the experiences of older LGBTQ+ individuals, particularly gay men, have identified several themes. Early in their lives there would be a recognition that they were attracted to people of the same sex, which could lead to sexual encounters.

Then for many of them there would be a period of denial where they would try to prove to themselves that they were in fact straight by getting married to someone of the opposite sex, a heterosexual marriage representing the ideal way of presenting oneself to the world at large and avoiding the criticism and stigmatisation of society. Denial represents a form of camouflage where people can present themselves as being something different to what they actually are and be seen as part of the majority. Communicating one's sexual orientation had to be done secretly and with some imagination.[86]

For some people the final stage will be, self-acceptance and recognition by others.[87] However, it is likely to be the case that far fewer older people are comfortable sharing their sexual orientation identity with others, which is borne out by the data.

The process of coming out for those of older generations was very stressful. They would experience discrimination which could have serious consequences in terms of health, well-being, career progress and income. One advantage of getting older for some people was that they were able to resist societal pressures to conform and reveal more of their true selves. It wasn't just in terms of the expression of their sexual orientation but more generally to reject the idea that growing older meant clinging onto youthfulness and agelessness.[88]

As described earlier, in his memoir, *The Glass Closet*, John Browne, now Lord John Browne, the former CEO of BP, wrote of how he felt about being gay. This was one of the most successful and powerful CEOs in the world and yet the word that crops up nearly 50 times in the book describing his experiences is fear or words related to it such as fearful and fearing. He questions whether if he had come out earlier in his career, he could ever have become a senior leader, never mind the CEO.

On average, older LGBTQ+ individuals experienced 6.5 lifetime victimisation and discrimination events—an "event" being something blatant and often aggressive including physical and verbal abuse, being threatened with being outed, property being attacked, being harassed by authorities, experiencing discrimination in the workplace and being forced out of areas they were living in. These are extremely significant and threatening forms of abuse to have suffered.[89] Lifetime victimisation and discrimination were also related to an increased likelihood of being disabled among LGBT older adults and of experiencing depression.

Furthermore because of the discrimination experienced at work, LGB adults were earning far less than the average for the population. Bisexual older people were far more likely to be concealing their identity, which led to a higher degree of internalised stigmatisation.[90]

Research carried out by Ashley looking at LGBTQ+ in organisations found that the changes in attitudes that have occurred over recent decades have had a considerable impact on older people who feel that it is easier for them to come out to the younger generation than to people of their own age. It leads to older people feeling much more accepted and far less likely to be abused.

As one participant said, "Younger generations are much more relaxed, so they don't care about this camp, middle-aged guy working with them, it's not relevant—they don't care . . . I've never had any comments here whereas previously people have sort of made jokes about it" (p174). This is revealing because it shows how much more valued and less judged this person feels today.

The change that has occurred and the way it impacts people of different ages are summed in this observation:

> you have some people who are like that guy was, who are in their 50s, who- when they started working in [the organisation] it was illegal- people lost their jobs over it and everything else . . . then you have 40s and people like me who are like "well I had a shit time at school and grew up in a shit town" . . . but if you talk to a 20 year old now, first of all they wouldn't identify just as gay, but I don't even think sexuality is at the forefront of their thoughts actually. (p175)

Issues and Obstacles Faced by Older LGBTQ+ People

Asexuality

There are very clear stereotypes about older people both positive and negative. The positive stereotypes include being dependable, reliable, committed to their work and less likely to quit compared to younger workers.

The negative stereotypes, however, include being less willing to change, not wanting to learn new skills, being less motivated in their work and consequently performing worse than younger people.

Research into the stereotypes, however, have shown the negative stereotypes to be unfounded but there is support for the positive ones.[91] As powerful and widely held as these attitudes are it's fair to say that they not only have very little validity, particularly the negative ones, but they get in the way of enabling us to see the positive contributions the older people bring to their work and to the workplace.

There are also powerful stereotypes about sexual attraction and what is considered appropriate for older people. This is summed up by a Kate participant in one study:

> I don't think it is portrayed in the media that's the thing. It's the little old lady and man little peck on the cheek. I don't think it's acknowledged that sexual desire goes on into older age or that it can be fun or interesting.[92]

Kate refers to the popular media and in movies and TV shows sex is physical, passionate and spontaneous. It's also the preserve of the young and the beautiful. In the workplace older people are stereotypically seen to be in decline, they are also seen as sexless or asexual, a perception that is shared with disabled people.

Portrayal of Beauty

The media portrayal of beauty being synonymous with youthfulness means that ageing for many people can lead to unrealistic expectations and frustration and dissatisfaction for failing to fulfil societal expectations. These feelings are exacerbated by the implicit assumption that to age well means retaining a youthful appearance and that this is somehow beneficial. Looks are given preference over enabling older people to find other ways of having a stimulating and satisfying life. Much of the focus on having an "ageless" body has been women but, with the increased emphasis on muscularity and leanness, it is an issue that impacts men including those in the older generations. Should someone feel that their body does not fit with current expectations of attractiveness, they will be less likely to engage in a sexual relationship.

However, the research shows that while societal expectations are particularly strong and negative, the actual experience of older people was varied, positive and self-affirming

There are considerable health and well-being benefits for older people having active sex lives. However, this health-related approach to sex of older people has led to a view that "successful ageing" requires older people to be engaged in sexual activity, and when they are not, it makes them feel as if they are responsible for the consequences.[93] The importance of taking a broader perspective is illustrated by a large-scale study from the United States which found that for older people it was the quality of the relationship rather than the quantity of sex, which was critical to health and well-being, something the researchers called sexual wisdom.[94]

A healthier approach is one in which there is acceptance of ageing at its core and then finding ways of maintaining a healthy and happy lifestyle. For those individuals who have moved away from the societal expectations of what a happy older age looks like they felt more empowered as a result, and their well-being was higher. This incorporates notions of bodily appearance as well as finding enjoyment from sexual activity in ways which recognise that conforming to stereotypes is particularly damaging.

The Invisibility of Bisexuality

When examining studies related to age and sexuality, heteronormativity is evident. Much of the research focuses on older heterosexuals with LGBTQ+ people noticeably absent. Where LGBTQ+ is considered, most attention is given to gay men.

The invisibility of bisexuality meant that there were few role models for them to identify with. One effect of this is that they were unable to be part of the community and to access the support which is a critical factor in helping to counter the stigmatisation.[95]

Self-stigmatisation

The stereotypes that are so widely held about older people then lead to not only them feeling stigmatised by others but also to being self-stigmatised. Individuals come to think of themselves as being no longer attractive and being "past it,"[96] which is something that impacts all people—heterosexuals and the LGBTQ+ community. However, this feeling of rejection is felt more severely by those who are LGBTQ+. This can mean that older people do not feel comfortable being in gay spaces in the way that they did when they were younger. This is partly due to an internalisation of negative attitudes which are felt more acutely by older gay men due to the promotion of youthfulness within the LGBTQ+ community.[97]

Acceptance

One key difference between older and younger members of the LGBTQ+ community is related to family relationships. For younger people the dilemma will be how to come out to their parents. For older members of the community the dilemma is reversed: how to come out to one's children and grandchildren.

A key consideration in coming out is the attitudes of society towards LGBTQ+ individuals: the higher the levels of homophobia, the greater the anxiety of coming out will be. Feel-good movies celebrate the coming out of older people, such as Beginners (which earned Christopher Plummer an Oscar for Best Performance by an Actor in a Supporting Role), but the truth is that the results of older people coming out to their families are far more mixed. Generally speaking, women were more active in ensuring that their families were accepting of their parents' sexual orientation than men were. Coming out to grandchildren was something that was particularly important to older LGBTQ+ individuals.[98]

Conclusion

For older generations of LGBTQ+ individuals it is harder for them to fulfil the principles of Identity Process Theory. There may be a realisation that keeping their sexual identity private was the most sensible and rational course of action given the discrimination experienced by those around them. There may be an acceptance that this is the way their life had to be lived, and no one can argue against that. However, there will be discontinuities in their lives, a reflection perhaps of what might have been and of opportunities missed. There will also have been impact in terms of their self-esteem and self-image. With societal attitudes towards LGBTQ+ changing positively in many parts of the world, we should find that more people of older generations will publicly share their sexual identities.

There are implications here in the way that organisations deal with creating the environment which is inclusive for older LGBTQ+ people. While for a significant number of people growing older means that they can express themselves more freely, there are others who because of the victimisation and discrimination that they will have experienced in their lives will still feel a sense of apprehension in seeking out support.

Furthermore, the stereotypes associated with older people (being over the hill, not attractive) will mean that some people will feel reluctant to become part of a community which they feel may judge them quite harshly. LGBTQ+ employee resource groups need to make sure that they don't perpetuate these norms in the way that they present themselves.

It is important that the ERGs are seen as being inclusive in everybody including age diversity.

There is much that can be gained by younger people by hearing about the experiences of older generations in adjusting what they expect from a relationship and the positive impact this will have on their sexual well-being.

Key points: Why Intersectionality Matters

Intersectionality is a powerful concept because it challenges our approaches to creating diverse and inclusive workplaces. More than that it challenges us all to examine our attitudes and biases, conscious and unconscious. Those from stigmatised groups (e.g., women, ethnic minorities) can sometimes feel, because they have experienced discrimination themselves, that this makes them, somehow, free from prejudice. It is a fanciful and deluded notion.

We started this chapter with Vice President Kamala Harris's declaration of love of Venn diagrams. We didn't give the whole quote, however, because she continued: "But here's what that Venn diagram also tells us: great, great, great reminder of the power of coalition building."[99] What we hope to have shown in this chapter is that the LGBTQ+ community has biases which creates exclusion, but by becoming more inclusive it can also become more influential.

Notes

1 www.youtube.com/watch?v=HVWwoCHJouU
2 Crenshaw, K.W. (1989). Demarginalizing the Intersection of Race and Sex: A Black Feminist Critique of Antidiscrimination Doctrine, Feminist Theory and Antiracist Politics. *University of Chicago Legal Forum*, 1989, pp. 139–167.
3 Leonard, W., & Mann, R. (2018). *The Everyday Experience of Lesbian, Gay, Bisexual, Transgender and Intersex (LGBTI) People Living with Disability, No.111 GLHV@ARCSHS*. Melbourne: La Trobe University, p. 8.
4 Breakwell, G.M. (1986). *Coping with Threatened Identities* (1st ed.). Psychology Press. https://doi.org/10.4324/9781315733913
5 Hirschmann, N.J. (2013). Queer/Fear: Disability, Sexuality, and the Other. *Journal of Medical Humanities*, 34(2), pp. 139–147.

6 De Than, C. (2015). Sex, Disability and Human Rights. In: T. Owens (ed.), *Supporting Disabled People with Their Sexual Lives*. London: Jessica Kingsley, pp. 86–103.
7 Assembly, U.G. (2006). Convention on the Rights of Persons with Disabilities. *Ga Res*, 61, p. 106.
8 www.who.int/teams/sexual-and-reproductive-health-and-research/key-areas-of-work/sexual-health/defining-sexual-health
9 Rich, A. (1981). *Compulsory Heterosexuality and Lesbian Existence*. London: Only women Press.
10 McRuer, R. (2006). *Crip Theory: Cultural Signs of Queerness and Disability*. New York: New York University Press.
11 www.nhsemployers.org/articles/understanding-and-supporting-staff-hidden-disability
12 NDA. (2005). *Disability and Sexual Orientation: A Discussion Paper*. Dublin: NDA. Accessed 20th June 2022.
13 Guzmán, P., & Platero, R.L. (2014). The Critical Intersections of Disability and Non-Normative Sexualities in Spain. *Annual Review of Critical Psychology*, 11, pp. 359–387.
14 NDA. (2005). *Disability and Sexual Orientation: A Discussion Paper*. Dublin: NDA. Accessed 20th June 2022.
15 NDA. (2005). *Disability and Sexual Orientation: A Discussion Paper*. Dublin: NDA, p. 13. Accessed 20th June 2022.
16 Trieschmann, R.B. (1988). *Spinal Cord Injuries: Psychological, Social, and Vocational Rehabilitation*. Demos Medical Publishing, p. 161.
17 Leonard, W., & Mann, R. (2018). *The Everyday Experience of Lesbian, Gay, Bisexual, Transgender and Intersex (LGBTI) People Living with Disability, No. 111 GLHV@ARCSHS*. Melbourne: La Trobe University.
18 Leonard, W., & Mann, R. (2018). *The Everyday Experience of Lesbian, Gay, Bisexual, Transgender and Intersex (LGBTI) People Living with Disability, No. 111 GLHV@ARCSHS*. Melbourne: La Trobe University.
19 Surfus, C.R. (2023). A Statistical Understanding of Disability in the LGBT Community. *Statistics and Public Policy (Philadelphia, PA)*, 10(1), p. 2.
20 Bollier, A.M., King, T., Austin, S.B., Shakespeare, T., Spittal, M., & Kavanagh, A. (2020). Does Sexual Orientation Vary between Disabled and Non-Disabled Men? Findings from a Population-Based Study of Men in Australia. *Disability & Society*, 35(10), pp. 1641–1659.
21 Hirschmann, N.J. (2013). Queer/Fear: Disability, Sexuality, and The Other. *Journal of Medical Humanities*, 34(2), pp. 139–147.
22 Tilley, E., Walmsley, J., Earle, S., & Atkinson, D. (2012). 'The Silence is Roaring': Sterilization, Reproductive Rights and Women with Intellectual Disabilities. *Disability & Society*, 27(3), pp. 413–426.
23 Sakellariou, D., & Algado, S.S. (2006). Sexuality and Disability: A Case of Occupational Injustice. *British Journal of Occupational Therapy*, 69(2), pp. 69–76.
24 Hirschmann, N.J. (2013). Queer/Fear: Disability, Sexuality, and the Other. *Journal of Medical Humanities*, 34(2), pp. 139–147.
25 Guzmán, P., & Platero, R.L. (2014). The Critical Intersections of Disability and Non-Normative Sexualities in Spain. *Annual Review of Critical Psychology*, 11, pp. 359–387.
26 Murphy, R. (1990). *The Body Silent*. New York: WW Norton.
27 Rohleder, P., & Swartz, L. (2009). Providing Sex Education to Persons with Learning Disabilities in the Era of HIV/AIDS: Tensions between Discourses of Human Rights and Restriction. *Journal of Health Psychology*, 14(4), pp. 601–610. https://doi.org/10.1177/1359105309103579

28 Wood, E., & Halder, N. (2014). Gender Disorders in Learning Disability—a Systematic Review. *Tizard Learning Disability*, 19(4), pp. 158–165.
29 McCann, E., Lee, R., & Brown, M. (2016). The Experiences and Support Needs of People with Intellectual Disabilities Who Identify as LGBT: A Review of the Literature. *Research in Developmental Disabilities*, 57, pp. 39–53.
30 Neufeldt, A. (2001). The Myth of Asexuality: A Survey of Social and Empirical Evidence. *Sexuality and Disability*, pp. 91–109.
31 Trieschmann, R.B. (1988). *Spinal Cord Injuries: Psychological, Social, and Vocational Rehabilitation*. Demos Medical Publishing, pp. 164–165.
32 Sakellariou, D., & Algado, S.S. (2006). Sexuality and Disability: A Case of Occupational Injustice. *British Journal of Occupational Therapy*, 69(2), pp. 69–76.
33 Macdonald, S.J., Donovan, C., & Clayton, J. (2017). The Disability Bias: Understanding the Context of Hate in Comparison with Other Minority Populations. *Disability & Society*, 32(4), pp. 483–499.
34 CREA. (2012). Count Me IN! Research Report on Violence Against Disabled, Lesbian, and Sex-Working Women in Bangladesh, India, and Nepal. *Reproductive Health Matters*, pp. 198–206.
35 Leonard, W., & Mann, R. (2018). *The Everyday Experience of Lesbian, Gay, Bisexual, Transgender and Intersex (LGBTI) People Living with Disability, No.111 GLHV@ARCSHS*. Melbourne: La Trobe University.
36 European Union Agency for Fundamental Rights. (2009). *Homophobia and Discrimination on Grounds of Sexual Orientation and Gender Identity in the EU Member States, Part II: The Social Situation*. EUAFR.
37 Kattari, S.K., Walls, N.E., & Speer, R.S. (2017). *Differences in Experiences of Discrimination in Accessing Social Services Among Transgender/Gender Nonconforming Individuals by (Dis)Ability. Journal of Social Work in Disability & Rehabilitation*, 16(2), pp. 116–140.
38 Lofgren-Martenson, L. (2009). The Invisibility of Young Homosexual Women and Men with Intellectual Disabilities. *Sexuality and Disability*, 27, pp. 21–26.
39 Stoffelen, J., Kok, G., Hosper, H., & Curfs, L.M. (2013). Homosexuality Among People with a Mild Intellectual Disability: An Explorative Study on the Lived Experiences of Homosexual People in the Netherlands with a Mild Intellectual Disability. *Journal of Intellectual Disability Research*, 57(3), pp. 257–267.
40 Wood, E., & Halder, N. (2014). Gender Disorders in Learning Disability—a Systematic Review. *Tizard Learning Disability*, 19(4), pp. 158–165.
41 Fredriksen-Goldsen, K.I., Kim, H.-J., Barkan, S.E., Muraco, A., Hoy-Ellis, C.P. (2013). Health Disparities Among Lesbian, Gay Male and Bisexual Older Adults: Results from a Population-Based Study. *American Journal of Public Health*, 103(10), pp. 1802–1809. https://doi.org/10.2105/AJPH.2012.301110
42 Wood, E., & Halder, N. (2014). Gender Disorders in Learning Disability—a Systematic Review. *Tizard Learning Disability*, 19(4), pp. 158–165.
43 Webb, M. (2014). *Accessing Canal Pride: The Intersection of Identities for LGBT People with Physical Disabilities at a Global Event*. Independent Study Project (ISP) Collection. https://digitalcollections.sit.edu/isp_collection/1983
44 Collective, C.R., Frazier, D., Smith, B., & Smith, B. (2017). *The Combahee River Collective Statement*. Gato Negro Ediciones.
45 Chulani, V.L., Barkley, L. et al. (eds.). (2019). *Promoting Health Equity Among Racially and Ethnically Diverse Adolescents*. https://doi.org/10.100 7/978-3-319-97205-3_12
46 Chulani, V.L., Barkley, L. et al. (eds.). (2019). *Promoting Health Equity Among Racially and Ethnically Diverse Adolescents*. https://doi.org/10.1007/978-3-319-97205-3_12

47 Grov, C., Bimbi, D.S., Nanín, J.E., & Parsons, J.T. (2006). Race, Ethnicity, Gender, and Generational Factors Associated with the Coming-Out Process Among Gay, Lesbian, and Bisexual Individuals. *Journal of Sex Research*, 43(2), pp. 115–121.

48 Chulani, V.L., Barkley, L. et al. (eds.). (2019). *Promoting Health Equity Among Racially and Ethnically Diverse Adolescents.* https://doi.org/10.100 7/978-3-319-97205-3_12

49 Grov, C., Bimbi, D.S., Nanín, J.E., & Parsons, J.T. (2006). Race, Ethnicity, Gender, and Generational Factors Associated with the Coming-Out Process Among Gay, Lesbian, and Bisexual Individuals. *Journal of Sex Research*, 43(2), pp. 115–121.

50 Mezey, N.J. (2008). The Privilege of Coming Out: Race, Class, and Lesbians' Mothering Decisions. *International Journal of Sociology of the Family*, pp. 257–276.

51 Kesslen, B. (2019, January 30). *NBC News.* www.nbcnews.com/feature/nbc-out/ lgbtq-people-color-face-compounded-violence-advocates-say-n964891

52 Meyer, I.H., Schwartz, S., & Frost, D.M. (2008). Social Patterning of Stress and Coping: Does Disadvantaged Social Statuses Confer More Stress and Fewer Coping Resources? *Social Science & Medicine*, 67(3), pp. 368–379.

53 Jaspal, R., & Siraj, A. (2011). Perceptions of 'Coming Out' Among British Muslim Gay Men. *Psychology & Sexuality*, 2(3), pp. 183–197.

54 www.stonewall.org.uk/sites/default/files/lgbt_in_britain_home_and_communities.pdf

55 Yip, A.K. (2004). Embracing Allah and Sexuality? South-Asian Non-Heterosexual Muslims in Britain. In: *South Asians in the Diaspora.* Brill, pp. 294–310.

56 www.stonewall.org.uk/system/files/rainbow_britain_report.pdf

57 Ayala, G., Bingham, T., Kim, J., Wheeler, D.P., & Millett, G.A. (2012). Modeling the Impact of Social Discrimination and Financial Hardship on the Sexual Risk of HIV Among Latino and Black Men Who have Sex with Men. *American Journal of Public Health*, 102(S2), pp. S242–S249.

58 Jaspal, R., Lopes, B., & Rehman, Z. (2021). A Structural Equation Model for Predicting Depressive Symptomatology in Black, Asian and Minority Ethnic Gay, Lesbian and Bisexual People in the UK. *Psychology & Sexuality*, 12(3), pp. 217–234.

59 Duran, A. (2021). "Outsiders in a Niche Group": Using Intersectionality to Examine Resilience for Queer Students of Color. *Journal of Diversity in Higher Education*, 14(2), p. 217

60 Khanolkar, A.R., Bolster, A., & Tabor, E., Frost, D.M., & Redclift, V. *Lived Experiences and their Consequences for Health in Sexual and Ethnic Minority Young Adults in the UK–A Qualitative Study.* London: University College London.

61 Khanolkar, A.R., Bolster, A., & Tabor, E., *Sexual and Ethnic Minority Young Adults in the UK–A Qualitative Study.*

62 Wade, R.M., & Harper, G.W. (2020). Racialized Sexual Discrimination (RSD) in the Age of Online Sexual Networking: Are Young Black Gay/Bisexual Men (YBGBM) at Elevated Risk for Adverse Psychological Health? *American Journal of Community Psychology*, 65(3–4), pp. 504–523.

63 Khanolkar, A.R., Bolster, A., & Tabor, E. Sexual and Ethnic Minority Young Adults in the UK–A Qualitative Study.

64 Alimahomed, S. (2010). Thinking Outside the Rainbow: Women of Color Redefining Queer Politics and Identity. *Social Identities*, 16(2), pp. 151–168.

65 Jaspal, R. (2017). Coping with Perceived Ethnic Prejudice on the Gay Scene. *Journal of LGBT Youth*, 14(2), pp. 172–190.

66 David, S., & Knight, B.G. (2008). Stress and Coping Among Gay Men: Age and Ethnic Differences. *Psychology and Aging*, 23, pp. 62–69.

67 Chulani, V.L., Barkley, L. et al. (eds.). (2019). *Promoting Health Equity Among Racially and Ethnically Diverse Adolescents.* https://doi.org/10.1007/978-3-319-97205-3_12

68 Chulani, V.L., Barkley, L. et al. (eds.). (2019). *Promoting Health Equity Among Racially and Ethnically Diverse Adolescents*. https://doi.org/10.1007/978-3-319-97205-3_12

69 Khanolkar, A.R., Bolster, A., & Tabor, E. Sexual and Ethnic Minority Young Adults in the UK–A Qualitative Study.

70 Lei, R.F., & Bodenhausen, G.V. (2017). Racial Assumptions Color the Mental Representation of Social Class. *Frontiers in Psychology*, 8, p. 519.

71 Johnson, K.L., Freeman, J.B., & Pauker, K. (2012). Race is Gendered: How Covarying Phenotypes and Stereotypes Bias Sex Categorization. *Journal of Personality and Social Psychology*, 102(1), p. 116.

72 Petsko, C.D., & Bodenhausen, G.V. (2019). Racial Stereotyping of Gay Men: Can a Minority Sexual Orientation Erase Race? *Journal of Experimental Social Psychology*, 83, pp. 37–54. https://doi.org/10.1016/j.jesp.2019.03.002

73 Agerström, J., Carlsson, M., & Strinić, A. (2021, November). Intersected Groups and Discriminatory Everyday Behavior: Evidence From a Lost Email Experiment. *Social Psychology*, 52(6), pp. 351–361. https://doi.org/10.1027/1864-9335/a000464

74 Devos, T., & Banaji, M.R. (2005). American=White? *Journal of Personality and Social Psychology*, 88(3), p. 447.

75 Pedulla, D.S. (2014). The Positive Consequences of Negative Stereotypes: Race, Sexual Orientation, and the Job Application Process. *Social Psychology Quarterly*, 77(1), pp. 75–94.

76 Johnson, J.D., & Ashburn-Nardo, L. (2014). Testing the "Black Code" Does Having White Close Friends Elicit Identity Denial and Decreased Empathy From Black In-Group Members? *Social Psychological and Personality Science*, 5(3), pp. 369–376.

77 Jaspal, R., Lopes, B., & Rehman, Z. (2021). A Structural Equation Model for Predicting Depressive Symptomatology in Black, Asian and Minority Ethnic Gay, Lesbian and Bisexual People in the UK. *Psychology & Sexuality*, 12(3), pp. 217–234.

78 Ryan, C., Russell, S.T., Huebner, D., Diaz, R., & Sanchez, J. (2010). Family Acceptance in Adolescence and the Health of LGBT Young Adults. *Journal of Child and Adolescent Psychiatric Nursing*, 23(4), pp. 205–213.

79 Del Río, C., & Alonso-Villar, O. (2019). Occupational Segregation by Sexual Orientation in the US: Exploring Its Economic Effects on Same-Sex Couples. *Review of Economics of the Household*, 17, pp. 439–467.

80 Del Rio, C., & Alonso-Villar, O. (2019). Occupational Achievements of Same-Sex Couples in the United States by Gender and Race. *Industrial Relations: A Journal of Economy and Society*, 58(4), pp. 704–731.

81 Webster, J., Thoroughgood, C., & Sawyer, K. (2018). Diversity Issues for an Aging Workforce: A Lifespan Intersectionality Approach. In: *Aging and Work in the 21st Century*. Routledge, pp. 34–58.

82 Rudolph, C.W. (2016, April). Lifespan Developmental Perspectives on Working: A Literature Review of Motivational Theories. *Work, Aging and Retirement*, 2(2), pp. 130–158. https://doi.org/10.1093/workar/waw012

83 www.ipsos.com/en-uk/ipsos-pride-survey-2024-gen-zers-most-likely-identify-lgbt

84 www.ons.gov.uk/peoplepopulationandcommunity/culturalidentity/sexuality/articles/sexualorientationageandsexenglandandwales/census2021#:~:text=The%20age%20group%20most%20likely,and%2034%20years%20(57.88%25)

85 https://news.gallup.com/poll/389792/lgbt-identification-ticks-up.aspx

86 Fredriksen-Goldsen, K., Hoy-Ellis, C., Kim, H.-J., Jung, H.H., Emlet, C.A., Johnson, I., & Goldsen, J. (2022a). Generational and Social Forces in the Life Events and Experiences of Lesbian and Gay Midlife and Older Adults Across the Iridescent Life Course. *Journal of Aging and Health*, 35(3–4), p. 089826432211255. https://doi.org/10.1177/08982643221125517

87 Neville, S., Kushner, B., & Adams, J. (2015, October) Coming Out Narratives of Older Gay Men Living in New Zealand. *Australasian Journal on Ageing*, 34(Supplement 2), pp. 29–33.
88 Towler, L.B., Graham, C.A., Bishop, F.L., & Hinchliff, S. (2021). Older Adults' Embodied Experiences of Aging and Their Perceptions of Societal Stigmas Toward Sexuality in Later Life. *Social Science & Medicine*, 287, p. 114355.
89 Fredriksen-Goldsen, K.I., Kim, H.-J., Shiu, C., Goldsen, J., & Emlet, C.A. (2014). Successful Aging Among LGBT Older Adults: Physical and Mental Health-Related Quality of Life by Age Group. *The Gerontologist*, 55(1), pp. 154–168. https://doi.org/10.1093/geront/gnu081
90 Fredriksen-Goldsen, K.I., Emlet, C.A., Kim, H.-J., Muraco, A., Erosheva, E.A., Goldsen, J., & Hoy-Ellis, C.P. (2012). The Physical and Mental Health of Lesbian, Gay Male, and Bisexual (LGB) Older Adults: The Role of Key Health Indicators and Risk and Protective Factors. *The Gerontologist*, 53(4), pp. 664–675. https://doi.org/10.1093/geront/gns12
91 Webster, J., Thoroughgood, C., & Sawyer, K. (2018). *Diversity Issues for an Aging Workforce: A Lifespan Intersectionality Approach*. In: *Aging and Work in the 21st Century*. Routledge, pp. 34–58.
92 Towler, L.B., Graham, C.A., Bishop, F.L., & Hinchliff, S. (2021). Older Adults' Embodied Experiences of Aging and Their Perceptions of Societal Stigmas Toward Sexuality in Later Life. *Social Science & Medicine*, 287, p. 114355.
93 Towler, L.B., Graham, C.A., Bishop, F.L., & Hinchliff, S. (2021). Older Adults' Embodied Experiences of Aging and Their Perceptions of Societal Stigmas Toward Sexuality in Later Life. *Social Science & Medicine*, 287, p. 114355.
94 Forbes, M.K., Eaton, N.R., & Krueger, R.F. (2017). Sexual Quality of Life and Aging: A Prospective Study of a Nationally Representative Sample. *The Journal of Sex Research*, 54(2), pp. 137–148.
95 Fredriksen-Goldsen, K.I., Shiu, C., Bryan, A.E.B., Goldsen, J., & Kim, H.-J. (2016). Health Equity and Aging of Bisexual Older Adults: Pathways of Risk and Resilience. *The Journals of Gerontology Series B: Psychological Sciences and Social Sciences*, 72(3), p. gbw120. https://doi.org/10.1093/geronb/gbw120
96 Towler, L.B., Graham, C.A., Bishop, F.L., & Hinchliff, S. (2021). Older Adults' Embodied Experiences of Aging and Their Perceptions of Societal Stigmas Toward Sexuality in Later Life. *Social Science & Medicine*, 287, p. 114355.
97 Towler, L.B., Graham, C.A., Bishop, F.L., & Hinchliff, S. (2021). Older Adults' Embodied Experiences of Aging and Their Perceptions of Societal Stigmas Toward Sexuality in Later Life. *Social Science & Medicine*, 287, p. 114355.
98 Orel, N.A. (2013b). Investigating the Needs and Concerns of Lesbian, Gay, Bisexual, and Transgender Older Adults: The Use of Qualitative and Quantitative Methodology. *Journal of Homosexuality*, 61(1), pp. 53–78. [Online]. https://doi.org/10.1080/00918369.2013.835236
99 www.youtube.com/watch?v=HVWwoCHJouU

Part III

Building an LGBTQ+ Inclusive Workplace

Chapter 8

The Evolving Workplace

Society continues to evolve in complex and often contradictory ways when it comes to LGBTQ+ issues. Navigating the effect on people's experiences within organisations is made all the more difficult by a self-reinforcing reluctance to manage the organisational response. In this chapter we explore three powerful factors that have led to the current conflicted state of LGBTQ+ inclusion in organisations: evolving legislation and policy, superficiality of action, and silence.

Legislation and Policy

Legal rights affecting LGBTQ+ people vary significantly across countries and jurisdictions, encompassing everything from anti-bullying legislation protecting LGBTQ+ children at school to the death penalty for homosexuality.

At the time of writing 66 jurisdictions criminalise private, consensual, same-sex sexual activity and almost half of these are Commonwealth jurisdictions.[1] Twelve countries retain the death penalty for private, consensual, same-sex sexual activity. At least six of these jurisdictions actually implement the death penalty—Iran, Northern Nigeria, Saudi Arabia, Somalia and Yemen—and the death penalty is a legal possibility in Afghanistan, Brunei, Mauritania, Pakistan, Qatar, UAE and Uganda. Fourteen countries criminalise the gender identity or expression of transgender people using so-called cross-dressing, impersonation and disguise laws. In many more countries transgender people are targeted by a range of laws that criminalise same-sex activity and vagrancy, hooliganism and public order offences. The result is that LGBTQ+ people can be lawfully arrested, detained and prosecuted in these places simply for being who they are. Even where these discriminatory laws are not enforced, their existence perpetuates stigma and prejudice towards the LGBTQ+ community through a lack of protection from discrimination and violence.

If you're reading this from the safety of an environment that does not criminalise the mere existence of LGBTQ+ people, then this situation may feel like an impossibility in your own circumstances. However, such laws

DOI: 10.4324/9781003489580-11

existed across the world in some form until relatively recently. For most of Europe the journey to decriminalisation of LGBTQ+ identities only began in 1967, when homosexuality was partially decriminalised in England and Wales (decriminalisation occurred earlier in Belgium, France and the Netherlands). The journey in the United States began later, in 1973. And journey is the right word in both cases because while significant progress has been made towards legal protection for LGBTQ+ people in these countries, it continues to be an ongoing process.

The significance of location and the norms and attitudes of the community are very important, with attitudes towards homosexuality becoming more positive in some parts of the world and gay marriage increasingly

Table 8.1 The global divide on acceptance of homosexuality

Ranking	Country	Acceptance (%)	Region
1.	Sweden	94	Europe
2.	Netherlands	92	Europe
3.	Spain	89	Europe
4=.	France	86	Europe
4=.	Germany	86	Europe
4=.	UK	86	Europe
7.	Canada	85	North America
8.	Australia	81	Oceania
9.	Argentina	76	South America
10.	Italy	75	Europe
11.	Philippines	73	Asia
12.	USA	72	North America
13.	Mexico	69	North America
14.	Japan	68	Asia
15.	Brazil	67	South America
16.	Czech Republic	59	Europe
17.	South Africa	54	Africa
18.	Greece	48	Europe
19.	Israel	47	Middle East
19.	Poland	47	Europe
21=.	Slovakia	44	Europe
21=.	South Korea	44	Asia
23.	India	37	Asia
24.	Bulgaria	32	Europe
25.	Lithuania	28	Europe
26.	Turkey	25	Middle East/Europe
27=.	Ukraine	14	Europe
27=.	Russia	14	Europe/Asia
27=.	Kenya	14	Africa
30.	Lebanon	13	Middle East
31.	Indonesia	9	Asia
32.	Nigeria	7	Africa

Adapted from "The global divide on homosexuality persists." Pew Research Center, Washington, D.C. (25 June 2020) www.pewresearch.org/global/2020/06/25/global-divide-on-homosexuality-persists/. Source: Spring 2019 Global Attitudes Survey. Q31.[4]

accepted. Where attitudes are becoming more positive, some of the negative impacts associated with being stigmatised are reducing, including LGBTQ+ health outcomes.[2] The highly respected Pew Research Organisation has produced a very helpful analysis of many countries in the world (see Table 8.1) which shows the acceptance of homosexuality.[3] There is a huge variation and unsurprisingly is related to legislation which either protects the LGBTQ+ community or allows discrimination against them. In North America, the percentage of people who said that homosexuality should be accepted by society was 72% in the United States and 85% in Canada. In Western Europe, the figures were uniformly high: 94% in Sweden, 92% for the Netherlands, 89% in Spain and 86% in France, Germany and the United Kingdom. The figures for other parts of Europe were markedly lower: 75% in Italy, 59% in the Czech Republic, with all other countries being below 50%. Outside of these regions, acceptance was highest in Australia (81%), Argentina, 76%, Philippines, 73%, Japan (68%) and Brazil (67%). In Nigeria, Tunisia and Indonesia the lowest acceptance scores were obtained (7%, 9%, and 9% respectively). In Russia only 14% agreed with the statement.

It is therefore no surprise that workplace anti-discrimination protection for LGBTQ+ employees also differs significantly across geographies. For example, in the United States, where LGBTQ+ identity has been decriminalised, there is no comprehensive federal law explicitly protecting LGBTQ+ individuals from discrimination and any protection largely depends on individual states enforcing local non-discrimination laws. Countries in Eastern Europe and parts of Asia have limited or no legal safeguard against discrimination. Countries in Western Europe on the other hand have strong and robust protection afforded by the Equality Act 2010 in the UK and the European Union's Employment Equality Directive. Younger readers in these countries may find it difficult to imagine a time when this protective legislation was not in existence. However, this legislation has only been enforced since 2003, and the damage that had already been done by that time continues to impact the experiences of LGBTQ+ people to this day.

Conducting research into the impact of protective legislation on LGBTQ+ employees is tricky. One of the reasons for this is that the lack of legislation makes it very difficult to recruit participants who will be open about their experiences.

Research comparing the existence and non-existence of legislation is typically conducted across different states in the United States. While these studies show a difference between states with and without legislation, there may be other factors contributing to these outcomes, such as differences in political and religious views which may impact community adoption of legislation.

Given these caveats it is fair to conclude that there are clear benefits to protective legislation including a reduction in recruitment discrimination towards LGBTQ+ applicants,[5] increased likelihood of LGBTQ+ friendly HR practices,[6] and improved organisational performance and economic benefits.[7]

Across Europe, the EU Labour Force Survey identified that the difference in unemployment rate is especially significant in countries with fewer equality and non-discrimination laws (5.9% versus 2.7%).[8] Recent findings show that legal relationship recognition is associated with statistically significant improvements in attitudes towards sexual minorities.[9]

Although there is a clear need for protective legislation for the LGBTQ+ community this need does not automatically translate into organisational policy. In 2002, 61% of Fortune 500 companies included sexual orientation in their corporate policies and 3% included gender identity.[10] By 2014, 91% prohibited discrimination based on sexual orientation and 61% on the basis of gender identity. In 2015, a review by LGBTQ+ network OUTstanding demonstrated that 80% of FTSE 100 firms' annual reports did not have specific non-discrimination policies for transgender staff.[11] Seventeen per cent of FTSE 100 company websites referred directly to transgender individuals in 2016.[12] Research carried out by INvolve in 2019 found that one-third of FTSE 100 company reports did not mention LGBTQ+ policies in their annual reports. This research also found that 17% of the FTSE 100 companies who had changed their Twitter logos for Pride Month did not mention LGBTQ+ in their annual reports.[13]

While these findings show an increase in the number of companies creating protective policies for LGBTQ+ staff, they also show that many of the largest UK organisations did not have a policy in place outlining their approach to LGBTQ+ discrimination and harassment—and the steps they were taking to prevent it. A Trades Union Congress (TUC) poll revealed that 21% of sampled organisations did not have a policy in place to support LGBTQ+ staff at work, and only 51% of managers surveyed told the TUC they had a policy prohibiting discrimination, bullying and harassment against LGBTQ+ workers in their workplace.[14]

You might expect that in an environment with appropriate legislation and comprehensive organisational policies to protect LGBTQ+ employees from discrimination, LGBTQ+ employees will be afforded safe and inclusive workplaces. Unfortunately, there's still much more work to do.

LGBTQ+ can be considered to be a subject that is often overlooked in organisations when it comes to diversity and inclusion. This became apparent to us when working with a large civil service department in the UK in 2017. This organisation was relatively advanced in its approach to LGBTQ+ inclusion with an anti-discrimination policy in place, an employee resource group up and running, and lanyards with pride flags visibly demonstrating employees' commitment to fairness and equality (before these were banned by the then Conservative government). The department had recently published their updated D&I strategy, the front cover of which proudly displayed a word cloud showing all the different protected characteristics that would feature in the strategy, including LGBTQ+. Sadly, this was the only mention of the acronym, and the identities encapsulated within it, in the entire document. It serves as a metaphor for many organisations' approach to LGBTQ+: showing the world that it is taking it seriously but nothing going on inside.

We have since spoken to many HR managers, D&I specialists, HR business partners, people partners and CEOs who have described their D&I strategy in the following way: "we're doing work on gender and ethnicity . . ." or "we have lots of employee resource groups, such as a women's network, an ethnic minority network . . ." Many of them are doubtless also concerned about LGBTQ+ inclusion and have employee resource groups to support LGBTQ+ employees, but there is a clear difference in the extent to which this topic is discussed and funded compared to other groups. Budgets available for LGBTQ+ inclusion support are extremely limited in comparison to other topics, and LGBTQ+ inclusion is usually placed within a broader learning and development or extra-curricular pot—and needs to be requested by employee resource group members at considerable expense and red tape.

The legislative landscape for LGBTQ+ people has been turbulent and inconsistent across jurisdictions so we cannot assume that all individuals are afforded the same protection. Where legislation exists, this does not automatically translate into action. In turn organisational policy does not automatically translate into active efforts to protect LGBTQ+ employees. Even in an organisation with a comprehensive strategy to improve LGBTQ+ diversity and inclusion, we can't rest on our laurels and assume that this will be enough to eradicate the challenges faced by the community. Implementing policies and legislation are necessary and important, but they do not immediately translate into changes in societal attitudes.

Silence

As practitioners and researchers in the topic of LGBTQ+ inclusion at work, we constantly hear that it's a subject on everyone's lips and undoubtedly going to be a huge issue for organisations . . . soon. Our experience working with organisations and studying the relevant research is that there is a very solid wall blocking the development of LGBTQ+ inclusion. Far from generating discussion and action, the topic seems to produce a discouraging silence.

We have seen how delays in protective legislation led to an oversight of LGBTQ+ inclusion by organisational policy, D&I managers and HR professionals in both formal and informal settings. But this delay isn't just a technical issue for managers and decision makers: it's harming people at the individual and team levels. There are several reasons for this reign of debilitating silence.

Let's not Talk about Sex

One D&I professional, responding to a verbal outline of some of our work with the LGBTQ+ community, cut in to tag the topic as one that "everybody's talking about" before rapidly segueing into dismissing it as irrelevant: "I don't even understand why this is an issue for organisations. . . . I don't care if my colleague is gay or straight, it's nothing to do with work—what people do outside of work is up to them, but there's no reason to bring it to work."

It is commonly thought that any discussion about sexuality is inappropriate for the workplace. However, this supposed standard is not enforced for heterosexual employees. The same D&I professional who refused to think about LGBTQ+ experiences had been happy enough to talk, unprompted, about his wife and children. This is an example of the double standard that is apparent when it comes to discussion of relationships at work.

One possible explanation for this double standard is a common conflation of a conversation about sexual identity with a conversation about sex. In the workplace, if a woman mentions her husband in passing or a man has a picture of his wife on his desk, this is not considered inappropriate. We don't conflate these behaviours with references to sex. However, if a woman mentions her wife in the workplace, or a man mentions his husband, that's a very different story.

The perception that disclosing non-heterosexuality at work is unprofessional also influences the behaviours of LGBTQ+ employees. A report from the Human Rights Campaign reported that half of the employees who had not disclosed their identity at work feared that doing so would make their colleagues feel uncomfortable and 25% felt that colleagues would consider their disclosure to be unprofessional.[15] A more recent report in 2018 by the Human Rights Campaign Foundation[16] in the United States further demonstrated the discomfort many employees feel when discussing LGBTQ+ topics and identities. While the report found that broad social acceptance for LGBTQ+ communities was at an all-time high, it also noted the pervasiveness of subtle biases. For example, the percentages of LGBTQ+ and heterosexual employees who felt comfortable discussing their spouse, partner or dating with co-workers were 73% and 78%, respectively. However, a quarter of employers reported that co-workers seemed uncomfortable when an LGBTQ+ colleague discussed their partner. Thirty-six per cent of heterosexual employees reported that they would feel uncomfortable hearing an LGBTQ+ colleague discuss dating, and 59% felt that it's unprofessional to discuss sexual orientation or gender identity at work at all.

The report found that 46% of heterosexual workers reported that they would not be very comfortable working with an LGBTQ+ colleague. The most common reason cited for this was that they "didn't want to hear about their co-workers' sex life."

At a public Pearn Kandola webinar in 2021, we discussed how LGBTQ+ identity can be seen as taboo in the workplace and explored the negative experiences this creates for members of the community. Afterwards we received an email from a client who wanted to anonymously share their experience. A decade previously they had been working for a large retail organisation and had been unexpectedly suspended for "bringing [their] inappropriate sex life into the workplace." Trying to learn what brought the suspension about it was nothing more than disclosing their bisexual identity to the team. They subsequently sued the organisation for unfair dismissal and won the case.

The Don't Ask, Don't Tell (DADT) policy enacted by the U.S. military further illustrates the pervasiveness of the perception that disclosure of non-heterosexuality is inappropriate for working environments. DADT was a controversial and historic U.S. military directive implemented in 1993 under the Clinton administration. The policy sought to address the longstanding ban on openly gay, lesbian and bisexual people serving in the armed forces by allowing non-heterosexual individuals to serve as long as they kept their sexual orientation a secret and did not speak openly about their identity. The impact of this was that lesbian, gay and bisexual service members were forced to live in secrecy and fear of being outed by others. Advocates of DADT argued that it would allow lesbian, gay and bisexual individuals to serve while maintaining the cohesion and morale of the military. However, over time it became clear that DADT was doing more harm than good, leading to the dismissal of thousands of skilled and dedicated military personnel and eroding trust and integrity within the armed forces. Consequently, DADT was repealed in 2010, marking a pivotal moment in the fight for LGBTQ+ rights and equality in the United States.

Lack of Awareness

A second factor contributing to the pervasive silence about LGBTQ+ identity in organisations is a general lack of awareness about the topic, which leads to diffidence in having conversations about it. This includes not understanding or knowing about LGBTQ+ identities, experiences and the issues faced in organisations.

Part of this can be attributed to the relative lack of exposure to LGBTQ+ individuals through media. GLAAD is an LGBTQ+ advocacy organisation that plays a significant role in monitoring and advocating for LGBTQ+ representation in the media, including television. As part of their work, they conduct an annual review of LGBTQ+ representation in scripted television programming across various broadcast and cable networks as well as streaming platforms. The most recent review published in 2023 showed that of the 659 series regulars in the 22–23 season, 70 of these characters (10.6%) were LGBTQ+. This is actually a decrease from the previous year where representation was 11.9%. Of those 70 LGBTQ+ characters, just seven were transgender characters. Of the 596 LGBTQ+ characters found across all platforms, there were 32 transgender characters counted in the report (5.4% of all LGBTQ+ characters). Of those there were 16 transgender women, 11 transgender men and 5 transgender nonbinary characters.

This marginalises transgender and non-binary experiences from our view. When we are exposed to LGBTQ+ characters, they are predominantly gay men or lesbians, which also marginalises bisexual individuals.

In addition to a lack of media representation, organisations have also missed opportunities to educate employees about this topic. The absence of LGBTQ+ employees in D&I policy and strategy for such a long time means that education and training on this topic has historically been scarce within organisations. In diversity and inclusion training, attention is often given to topics such as gender, ethnicity and disability but not to the experiences of LGBTQ+ employees.

An additional challenge for organisations—and one that stems directly from delayed implementation of legislation and policy—is a lack of data about LGBTQ+ employees. The LGBTQ+ community has been referred to as the largest but least studied minority group in the workplace.[17,18] Businesses seldom collect data about LGBTQ+ identity and employee experience. When this data is collected it is often aggregated into the experiences of one group as a whole—yet we know the experiences of individuals within this umbrella community are vastly different.

The lack of research and organisational data on this topic means that the challenges faced by LGBTQ+ employees are hidden from the leaders who can implement changes. As the issue is not quantified, it is much easier to overlook. In fact, not collecting data about employee experiences could have a more pronounced impact on the LGBTQ+ community than on other groups: the invisibility of this identity means we cannot see issues of discrimination across organisations in the same way that we can see issues of gender and race. Without the data we could be led to believe that LGBTQ+ inclusion is not such an issue in an organisation.

Language

The third explanation for the relative silence about this topic is the rapid evolution of language relating to the LGBTQ+ community. The terms and labels ascribed to individuals within the LGBTQ+ community are as broad as the community itself. Changes to the acronym have reflected society's developing understanding and recognition of different identities. It is for this reason that variations of the acronym are widespread, including LGBT, LGBTQ, LGBTQ+ and LGBTQIA+. It is important to recognise, however, that this isn't a new development; the language and terminology have been changing ever since Kertbeny coined the terms homosexual and heterosexual.

In addition to the growing recognition of different identities within this community, some terminology has become outdated. This has again occurred due to growing knowledge of the topic and is likely to continue happening as people become more aware. On the other hand, some terms that were once considered to be derogatory and were used in a pejorative way have since been reclaimed by LGBTQ+ individuals. One such example is the term "Queer." Younger people use this term more freely than older generations who have memories of "queer" always being used as an insult and to cause hurt.

Changes in language and fear of using terms that are outdated lead people to avoid engaging in conversations about LGBTQ+ identities and inclusion. This avoidance is a way of protecting oneself from being penalised for causing offence and also protecting others from such offence. However, staying silent or avoiding discussion can itself send a signal to others about the extent to which we value them. The other key impact of this avoidance is that people are likely to miss opportunities to learn more and become more confident in these discussions by avoiding informal conversation and exposure.

The Cycle of Silence

The cycle of silence refers to how a limited discourse on sexuality in organisations perpetuates a lack of understanding, which, in turn, means that no action is taken to address the issues that the LGBTQ community face at work.

The significant lack of discussion in organisations about LGBTQ+ identities and inclusion has a knock-on effect. If we avoid speaking with LGBTQ+ colleagues because we're scared that we'll say the wrong thing, then we miss opportunities to learn the right thing to say. If we don't collect data about LGBTQ+ employees' experiences, then we simply won't know about how safe and included they feel at work or what support they need. If we don't ask about what is being done to support LGBTQ+ colleagues because we don't know enough about what should be done, then we miss opportunities to implement training programmes to plug the gap. And when we miss these opportunities, we're perpetuating the lack of understanding.

Further, if we don't understand the challenges that colleagues are facing then we don't know what needs to change. The issue remains hidden from those who can implement change and the situation doesn't improve.

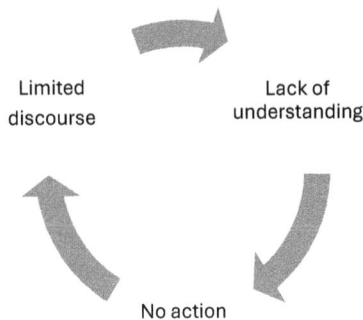

Figure 8.1 The Cycle of Silence

Policy can't be designed to help overcome these issues and no action is taken. LGBTQ+ networks and employee resource groups will not be created and resourced.

Not taking action, and not visibly committing to supporting LGBTQ+ colleagues, creates an environment where LGBTQ+ employees don't feel safe to disclose their identity and others don't feel comfortable to bring up the topic. Ultimately, everybody remains silent.

This cycle is self-perpetuating. Silence perpetuates the issue by reinforcing the idea that this is a taboo subject, minimising opportunities for education and preventing organisations from developing support and resources.

Silence about LGBTQ+ identities and challenges has become the norm. Several cues in the organisational environment contribute to this effect include:

- referring to some aspects of diversity and inclusion, such as gender, ethnic minorities, but not talking directly about LGBTQ+;
- avoiding discussion of LGBTQ+ identity;
- excluding LGBTQ+ colleagues from issues of concern to them;
- shutting down conversations when LGBTQ+ identities are raised.

Due to the power of social norms, silence can become in danger of becoming impenetrable.

Action

Start Somewhere

Pride Month is notable for the increased number of rainbow flags and logos on display both in the real world and online. Companies are keen to display their credentials when it comes to LGBTQ+ inclusion in their workplaces and to demonstrate their commitment to ensuring that their organisations enable everyone to bring their true selves to work. The question that can legitimately be asked of course is what happens for the other 11 months of the year? The members of the public remain to be convinced that this represents a true change in organisations' attitudes with more people believing that organisations do this to gain good publicity rather than genuinely wanting to do the right thing.[19] The gap between legislation being enacted and organisations developing comprehensive policies is also reflected in the motivations for engaging in Pride Month. On the one hand, organisations will be genuinely interested in showing their commitment to LGBTQ+ inclusion, but there is also clearly a commercial benefit to them by doing this.

It's worth sharing another point of view on this.

The Commonwealth Games is one of the biggest multi-games event in the world, second only to the Olympic games. The 2014 edition was hosted in Glasgow, Scotland, where nearly 5,000 athletes competed in 17 different sports. The opening ceremony of the Commonwealth Games always suffers in comparison with that of the Olympic Games, with the budget being far smaller. Glasgow's ceremony was colourful, humorous and featured dancing Tunnock's chocolate tea cakes. Towards the end of the ceremony, the actor John Barrowman briefly kissed one of the male dancers on the lips. For many people in the UK, this was of no particular significance. However, in 42 of the competing countries, homosexuality was still a crime. The brief contact wasn't intended only for the home audience but was sending a message to those countries that their views were not universally held. To LGBTQ+ people in those countries it would have let them know that they were not alone and nor were they abnormal.

We should understand that there are different ways of being activist. The more muscular, overtly demonstrative form is the one that is generally seen as being the most acceptable to some. We should also recognise, however, that, as mild as it may seem, even wearing a little rainbow lanyard will not only demonstrate a level of support for LGBTQ+ inclusion but also be seen as provocative by some in other parts of the world. It is remarkable the power that the rainbow symbol has because of what it represents. Qatar, UAE, Turkey, Hungary and China are some of the countries that have banned the rainbow symbol in any form. When organisations, particularly global ones, demonstrate their support for Pride Month, we need to appreciate that this will be sending a message to their employees in countries which hold highly homophobic attitudes that they have support from colleagues in other regions.

As superficial as these gestures may seem, they do represent a significant change from only 20 years ago and that should be celebrated, while at the same time we need to recognise that more can and should be done.

Standards and Awards

The accusation can also be made of organisations that they seek to achieve awards to display their commitment to action but that nothing really changes within the organisation itself. The standards, and there are an increasing number of them, that seek to reward organisations for the inclusion of LGBTQ+ staff are often quite superficial in the assessments that they make. The standards themselves have helped to bring about an awareness but they don't necessarily bring about more change than that. Therefore the organisation can take credit for being on a list of like-minded organisations, but the awards process itself actually limits creativity and can lead to an institutionalisation of ineffective practice. It is important therefore the organisations seek to take action that are based on evidence and which are relevant to the context in which they are operating. This will enable them to not just take a formulaic, tick-box approach to the issue but instead approach this with a sense of flexibility and innovation.

The Power of Social Norms

Our beliefs, attitudes and experiences all have an impact on our behaviour. The most powerful influence, however, is social norms—the unwritten rules of beliefs, attitudes and behaviours that are considered acceptable in a particular group or culture. Norms provide us with an expected idea of how to behave, and they function to provide order and predictability in society. Every time a behaviour happens, it becomes more and more normalised.

For example, when people staying at a hotel are told that the majority of other guests there reuse their towels, they are significantly more likely to do this themselves compared to simply being told this behaviour is preferable for environmental reasons.[20] Studies have also demonstrated that awareness of neighbours' recycling behaviours significantly increases the likelihood that somebody will themselves recycle.[21] This has become such a powerful social norm that people who don't recycle are often considered to be selfish and anti-social. We also know that people are more likely to drive under the influence if it is seen as a normal behaviour among their peer group or community.[22] It is important therefore that we able to discuss issues relating to LGBTQ+ openly and honestly in organisations. It also means that we create safety for all people to express their views particularly on more contentious topics. It also means that we can provide feedback to one another on where we may have caused hurt or offence to others unintentionally.

Key Points

Organisations are having to catch up and ensure that their diversity and inclusion policies are genuinely inclusive of LGBTQ+ communities. We feel that organisations generally speaking, while they have good intentions, have been slow to seriously consider LGBTQ+ issues in the workplace.

A combination of factors including lack of awareness, limited conversation and concerns about the correct terminology to use are some of the factors which contribute to silence on this issue.

There has been criticism of some organisations only taking symbolic action for example during Pride Month. We have a more sympathetic view on this and recognise that even something as simple as displaying a rainbow flag on your website will demonstrate your support and will send a message to homophobic individuals and regions that their views are not acceptable. This is a significant change that has happened over the 20 years.

It's also important that organisations continue to make progress and to ensure that the LGBTQ+ inclusion becomes a reality. More of the actions that can be taken will be described in the next chapter.

Notes

1 humandignitytrust.org
2 Mustanski, B., Birkett, M., Greene, G.J., Hatzenbuehler, M.L., Newcomb, M.E. (2014). Envisioning an America without Sexual Orientation Inequities in Adolescent Health. *American Journal of Public Health*, 104(2), pp. 218–225.

3 Poushter, J., & Kent, N. (2020). *The Global Divide on Homosexuality Persists.* Pew Research Center, p. 25.
4 Poushter, J., & Kent, N. (2020). *The Global Divide on Homosexuality Persists.* Pew Research Center, p. 25.
5 Tilcsik, A. (2011). Pride and Prejudice: Employment Discrimination Against Openly Gay Men in the United States. *American Journal of Sociology*, 117(2), pp. 586–626.
6 Everly, B.A., & Schwarz, J.L. (2015). Predictors of the Adoption of LGBT-Friendly HR Policies. *Human Resource Management*, 54, pp. 367–384. https://doi.org/10.1002/hrm.21622
7 Hossain, M., Atif, M. et al. (2020). Do LGBT Workplace Diversity Policies Create Value for Firms? *Journal of Business Ethics*, 167, pp. 775–791. https://doi.org/10.1007/s10551-019-04158-z
8 ILGA-Europe Rating of EU Labour Force Survey. https://rainbow-europe.org/country-ranking
9 Aksoy, C.G., Carpenter, C.S., De Haas, R., & Tran, K.D. (2020). Do Laws Shape Attitudes? Evidence from Same-Sex Relationship Recognition Policies in Europe. *European Economic Review*, 124, p. 103399. https://doi.org/10.1016/j.euroecorev.2020.103399
10 Human Rights Campaign, Corporate Equality Index. (2002). https://assets2.hrc.org/files/assets/resources/CorporateEqualityIndex_2002.pdf
11 Bentley. (2015). *OUTstanding Review.* www.huffingtonpost.co.uk/neil-bentley/businesses-lgbt- inclusion_b_8004434.html?utm_hp_ref=tw
12 Beauregard, T.A., Arevshatian, L., Booth, J.E., & Whittle, S. (2016). Listen Carefully: Transgender Voices in the Workplace. *International Journal of Human Resource Management.* https://doi.org/10.1080/09585192.2016.1234503
13 www.thehrdirector.com/business-news/diversity-and-equality-inclusion/one-three-ftse-100-dont-mention-lgbt-annual-reports-1532019/
14 www.tuc.org.uk/news/1-5-workplaces-do-not-have-any-policies-support-lgbt-staff-tuc-poll#:~:text=Bullying%20and%20harassment%3A%20Only%20half,LGBT%20workers%20in%20their%20workplace
15 2009 Human Rights Campaign: Degrees of Equality. https://assets2.hrc.org/files/assets/resources/DegreesOfEquality_2009.pdf
16 2018 Human Rights Campaign: A Workplace Divided. https://hrc-prod-requests.s3-us-west-2.amazonaws.com/files/assets/resources/AWorkplaceDivided-2018.pdf
17 Ragins, B.R. (2004). Sexual Orientation in the Workplace: The Unique Work and Career Experiences of Gay, Lesbian and Bisexual Workers. In: *Research in Personnel and Human Resources Management.* New York, NY: JAI Press, pp. 35–129.
18 Ozeren, E. (2014). Sexual Orientation Discrimination in the Workplace: A Systematic Review of Literature. *Procedia—Social and Behavioral Sciences*, 109, pp. 1203–1215. https://doi.org/10.1016/j.sbspro.2013.12.613
19 Sánchez-Soriano, J.J., & García-Jiménez, L. (2020). The Media Construction of LGBT+ Characters in Hollywood Blockbuster Movies. The Use of Pinkwashing and Queerbaiting. *Revista latina de comunicación social*, (77), pp. 95–115.
20 Goldstein, N.J., Cialdini, R.B., & Griskevicius, V. (2008). A Room with a Viewpoint: Using Social Norms to Motivate Environmental Conservation in Hotels. *Journal of Consumer Research*, 35, pp. 472–482. https://doi.org/10.1086/586910
21 Thomas, C., & Sharp, V. (2013). Understanding the Normalisation of Recycling Behaviour and Its Implications for Other Pro-Environmental Behaviours: A Review of Social Norms and Recycling. *Resources, Conservation and Recycling*, 79, pp. 11–20,
22 González-Iglesias, B., Gómez-Fraguela, J.A., & Sobral, J. (2015). Potential Determinants of Drink Driving in Young Adults. *Traffic Injury Prevention*, 16(4), pp. 345–352. https://doi.org/10.1080/15389588.2014.946500

Chapter 9

Organisational Actions to be LGBTQ+ Inclusive

What should organisations do to create an environment where LGBTQ+ individuals don't just feel "tolerated," or "accepted," but are able to thrive and contribute to their full potential?

In this chapter we will explore the actions that will support organisations in achieving an inclusive environment, including implementing and maintaining a comprehensive inclusion policy, providing education for employees, collecting sufficient data and using this to drive initiatives, supporting employee resource groups and networks, and ensuring ongoing visibility of support.

Organisational Policy

For an LGBTQ+ inclusion policy to be effective, there are five key elements that we need to have.

First, you need a policy. This might sound obvious, but actually LGBTQ+ employees are often overlooked in organisational policy and strategy. Having a policy is an essential starting point.

Second, the policy recognises heterogeneity. Research consistently demonstrates that the experiences of lesbians, gay men, bisexual men and bisexual women are all different, which are again different to the experiences of transgender and non-binary individuals (the experiences of binary and non-binary transgender employees are again very different).

To give you an example, bisexual individuals often face greater levels of bullying and prejudice than lesbian and gay individuals[1] and transgender employees are more likely to conceal their identity than other sexual minorities.[2] In addition, some members of the transgender community may experience medical or surgical transition, which is not an experience that is shared by other members of the LGBTQ+ community. By assuming that the LGBTQ+ community share many experiences, we can easily overlook these details in organisational policy, which means that the voices of those most severely affected by discrimination and exclusion remain unheard.

Third, the policy recognises barriers for LGBTQ+ employees in reporting negative behaviours through traditional reporting mechanisms, which can

DOI: 10.4324/9781003489580-12

mean individuals having to disclose their sexuality or gender identity to make a complaint. Such disclosure can potentially expose them to further discrimination or exclusion, and so such experiences will often go unreported.

One way of reviewing the current reporting mechanism or disciplinary procedure in your organisation is to look at it from the perspective of someone with an LGBTQ+ identity but is fearful about disclosing it. How would they navigate this process and how might they feel? It is worth remembering that 65% of LGBTQ+ employees who had been harassed did not report it.[3]

Things that can help here are the implementation of anonymous reporting mechanisms and also having support systems for the LGBTQ+ community that do not require people to disclose their identity or asking their line manager.

Fourth, ensure that other policies are inclusive and that they don't overlook the experiences of LGBTQ+ employees, for example, employee benefit policies. This is important in and of itself, and especially so as 77% of LGBTQ+ job seekers consider it an indicator of the company's commitment to LGBTQ+ inclusion[4]. Organisations may also want to look beyond adapting existing benefits and also incorporate specific and targeted benefits such as adoption assistance, parental leave for both partners, bereavement leave and transgender-specific medical coverage. Where organisations do offer such specific benefits, employees are significantly more likely to be open about their identity.[5]

Fifth, employees know about it! Employees need to be aware of the protection, support and assistance that is available to them, so ensure this is well distributed and communicated, both to existing staff and also new joiners.

As we mentioned in Chapter 1, when it comes to issues relating to gender identity organisations will need to ensure that they do not promote the interests of one group over another. Whilst debates on this topic can become toxic, the evidence for what constitutes best practice remains weak. It is important that we recognise the need to ensure that different points of view can be heard in an environment that feels safe to do so, otherwise we run the risk of policy being determined by those who shout the loudest.

Reviewing Policies

Existing policies also need to be reviewed particularly those related to human resource practices.

It is essential that policies covering the employee life cycle (from recruitment through to retirement) are looked at through an LGBTQ+ lens to ensure that there is fair and effective as they can be, in line with the current best practice.

Those involved in managing and implementing the policies and processes also need to be trained in them. Too often, issues relating to bias against the LGBTQ+ community are overlooked and ignored in such training. It can leave the appearance that it's not important and indeed that such bias no longer exists.

Education

Throughout this book we have identified a lack of awareness among employees, HR professionals and leaders about the issues faced by LGBTQ+ colleagues, and also about recognising this area of identity when thinking about diversity and inclusion. Education is key to ensuring that individuals feel empowered and comfortable to have conversations about LGBTQ+ identities and inclusion, and it is therefore important that organisations understand their duty to educate employees on this topic.

Education can take various forms: from training and workshops to reverse mentoring of leaders. Such initiatives open dialogue between people and done well, provide opportunities to learn in a safe environment and see what it's really like to be an LGBTQ+ individual in your organisation.

It makes sense to involve and engage your LGBTQ+ employee resource group where they are willing and able, but individuals should not be expected to put themselves in a position where they may risk exposing themselves to discrimination and exclusion.

Training should also cover other areas as well for example in how to be an active bystander.

People are more likely to intervene when someone behaves in an overtly objectionable way.

The more subtle the behaviour the less likely it is that anyone will intervene, but they are also less likely to have identified it as being offensive. It is really important therefore that awareness is raised of micro-incivilities and the impact these have on individuals. When someone behaves in an overtly discriminatory way, it is clearly going to be stressful. However, the perpetrator will be identifiable. With the micro-incivilities, there will be ambiguity in terms of interpreting the behaviour: Did they mean it? Was it intentional? How do you respond? How did others react? As a result of these questions, individuals are far less likely to take any action but will continue to relive the moment, trying to understand what happened. It is this long-term rumination that causes the stress especially when these behaviours are repeated on a regular basis. As a consequence, it will have a long-term effect on people's psychological well-being if they are in an environment where these types of behaviours occur regularly.

Training shouldn't just be on the specific topic of LGBTQ+ but it needs to be incorporated into other forms of training for example inclusive leadership. But beyond that into training for interviewers, on performance management, promotion systems and so on. Where staff are being given skills training or teambuilding, then you need to ensure that this is referred to.

There are other ways we can all increase our awareness of these topics, for example, by joining resource groups as allies, supporting others and taking opportunities to learn about their experiences. Diversifying the media we are exposed to is a great way to increase knowledge and there are helpful resources online where we can continue our own learning. It also includes reading novels about the lived experiences of individuals with a different identity to us, as this is a way of increasing our empathy towards them.[6]

Data

Data is extremely important when talking about diversity and inclusion, because it highlights progress as well as challenges. However, we know that LGBT experiences are not just under-represented in the research literature but also in organisational data.

Being LGBTQ+ is an identity that is invisible and so having data, it could be argued, is more important for this community. It also presents practical problems in how to collect the information in a way that is safe and confidential. There are a few things that you need to consider when doing this.

Where?

As we saw in Chapter 8, anti-discrimination legislation varies significantly across geographies, and such laws extend to the type of information organisations are able to collect from employees. Ensure that you are aware of the legislation in the relevant countries or states, as this will impact the type of data you can ask for as well as the way you are allowed to collect it.

Who?

Think about who you are hoping to collect data from. If you're trying to gain an understanding about the experiences of those within the LGBTQ+ community, you should make this inclusive of all within it. Make sure that the language you're using to collect the data is inclusive, because restrictive, or binary identity categories may signal to employees that you haven't given enough consideration to the individuals you are requesting information from. Engage your LGBT network in the design of this data collection, to ensure that the language you're using and the questions you're asking are appropriate.

Ideally, using a mixed-method approach is best. This will be a combination of quantitative (e.g., surveys) and qualitative (e.g., interviews, focus groups).

Once you have survey data, it's important, in the first instance at least, that it's not aggregated into one single set. Aggregating data across the community makes it much less meaningful and doesn't give a clear picture of unique experiences across the population. Sample sizes mean that you may have to do this, but it is important to acknowledge and be transparent about the aggregation of data and the limitations this has on interpretation. Qualitative data is often used in research as this enables individual voices to be heard.

What?

When measuring LGBTQ+ inclusivity, employers also need to consider what type of data are they looking for.

There's a tendency for employers to focus on extreme forms of negative behaviours, such as discrimination, homophobia, biphobia and transphobia. These behaviours do occur, but LGBTQ+ employees are less likely to report it as explicit discrimination, as there's a reluctance to believe that these experiences have been driven by such negative motivations.

By focusing on such explicit forms of negative behaviour, employers may be overlooking modern or implicit forms of discrimination and exclusion. Some of the most common behaviours that we often hear about from LGBTQ+ employees, and some of the hardest to manage, are those that are somewhat ambiguous and where the intent to harm is less obvious. This could include somebody asking intrusive questions about their private life or using incorrect gender pronouns.

Gathering data on the full spectrum of experiences will allow an employer to get a true sense of how their LGBTQ+ employees are being treated in the workplace and just how included they feel.

How?

Next employers need to consider how they're going to collect this information. Anonymity and confidentiality are of the utmost importance, especially if there's a small sample. In some cases, there may only be one employee within an organisation who identifies with a certain LGBTQ+ identity. Employers need to think this through carefully and identify where anonymity might be threatened and how employees will be protected.

At all stages of data gathering, employers should be transparent and straightforward with their employees about how they're going to use this data. This will increase the chances of employees providing feedback in the future and being candid with their feedback.

When planning the method of data collection, employers should consider the different ways to do this. Surveys can be a convenient way to collect a lot of data quickly, and they can allow for anonymity, but it's often difficult to identify what questions should be included. For this reason, some initial interviews or focus groups at the beginning of the process can be effective in identifying the behaviours or experiences to focus on.

Employers might want to consider bringing in an external party to run their data collection, as focus groups require a high degree of disclosure from participants. If the person running these groups is internal, employees may be reluctant to speak up. In the past we have found that employees talk more openly to us because we don't know them or their colleagues and so the risk of disclosing their identity is minimised. It also demonstrates a commitment from the organisation to understanding their experiences.

You might find that your first attempt at data collection does not attract a large sample, but consistency and transparency in demonstrating how you're using the results to make effective change will increase participation in the future. Not being transparent nor following up in a meaningful way will mean LGBTQ+ employees are unlikely to feel motivated to share their experiences with you in the future.

Networks and Employee Resource Groups

Deloitte, one of the big four accountancy and consultancy organisations, announced in the spring of 2017 that they would be getting rid of their employee resource groups and replacing them with inclusion councils. The response to the decision was mixed: some criticise the firm, while others agreed that such groups don't help organisations to achieve greater inclusion.[7] PWC, one of Deloitte's principal competitors, took the opportunity to reaffirm their commitment to such groups, seeing them as an important part of their plans.[8] Generally speaking, we don't view ERGs as a problem or as the solution—instead, they are a symptom of the lack of inclusion that exists in organisations. Groups exist to provide a safe space for employees to discuss the issues that impact them and provide support to one another.

When it comes to the LGBTQ+ community, however, it is important to recognise the significance of having such networks. The history of sexual minorities, particularly since the 18th century, has been one of increasing isolation and secrecy. As we saw in earlier chapters, there has been a need for individuals to meet up in safe locations where they can be their true selves. Whether this is in the Molly houses of this 18th century all the Mattachine Society of the 20th century, the need to congregate together, to feel secure and to even advocate for changes in society has been an important part of people's lives. Establishing networks in organisations isn't the new initiative that we might think it is; rather, it is an extension into the workplace of a centuries-old practice.

LGBTQ+ networks (or sometimes known as Employee Resource Groups) are a significant way for organisations to demonstrate support. Many organisations have networks in place; however, they are not all supported to the same extent.

More often than not, such networks are grassroots generated and rely on the dedication of members, who give up their time with no additional support or even recognition for the work that goes into the development of the network.

Research, conducted within the National Health Service, the UK's largest employer, showed that LGBTQ+ network chairs were having to complete network-related tasks on top of their normal working hours.[9] The risk is that over time this potentially valuable resource becomes unsupported and loses its effectiveness.

If you don't already have one, consider setting up an LGBTQ+ network that is well supported by leadership, giving the leaders of the network the formalised time allocation and support they need to build it, engage employees and create a safe space. If you are a small organisation, support your staff in joining external LGBTQ+ networks, which are often set up to support those across industries.

The resource group also needs to ensure that it is genuinely inclusive. As we have seen, some groups of people can feel isolated and lonely because they are not only rejected by their family and friends but also by the LGBTQ+ community. The network can have its own awareness sessions to understand the issues faced by for example LGBTQ+ minorities, disabled people and older people.

The group can then be a safe space for all individuals to express their concerns and to experience empathy from them. This sensitivity will extend to people appreciating that actions that seem appropriate in the majority community may not be exactly the same for those in minorities. For example, the process of coming out for minorities isn't necessarily the same for those from the majority community.[10]

Allies

Allies can be an incredibly powerful in reducing the discrimination and exclusion that employees face, as they are more likely to intervene and be an active bystander in those situations. The heterosexual people who are most likely to become allies to the LGBTQ+ community have a particular profile. They are more likely to be:

- female
- white
- politically more liberal and left-leaning
- less actively involved in religion
- concerned about other forms of exclusion for example on the grounds of gender and race
- involved because of discrimination they have seen experienced by LGBTQ+ friends and acquaintances
- comfortable in their relationships with LGBTQ+ friends

They may also hold "positive" stereotypes about gay men in particular and will also be aware of being stigmatised by association.[11]

There are implications from understanding the profile of an ally. First, it emphasises the need to ensure that LGBTQ+ inclusion is discussed in organisations as part of the broader discussions that take place about gender and race. For those individuals who have already engaged in making workplaces more inclusive on those grounds, understanding these connections will motivate them to look at other aspects of identity. Second, while allies are less likely to support negative stereotypes, they also need to be made aware of the dangers of holding positive stereotypes. Third, allies need to be supported in how to deal with stigmatisation by association.

By creating these links and providing support, more allies can be encouraged to get involved.

Finally, ensure that your network feels welcoming to allies to join. If they sense that this is not for them, that will limit the support that they feel that they can offer.

Establishing Norms

People often talk rather abstractly and vaguely when they refer to culture change. A change in the culture occurs where the accepted norms are questioned, challenged and new behaviours are adopted. There are many theories about behaviour change, but one of the most influential is the theory of planned behaviour.[12] Essentially to bring about the change three questions need to be answered:

1. What's in it for me?
2. What will the reaction be of others if I don't change?
3. How difficult is it?

The answer to the first question for many organisations when it comes to DEI-related topics is to look at money—finding a way of rewarding people for engaging in inclusive behaviours (which often relate to the degree of diversity you have in your team) or of punishing them by not rewarding them so much. What is underestimated is the extent to which people are motivated to do the right thing. Reminding people that they are fair, seek to treat people with respect and want to create an environment where people can give their best is a powerful way of demonstrating the benefits namely that our behaviour will be in line with our values and self-image.

In addition to this we can explain to people the business benefits of creating a more inclusive environment for everyone and for members of the LGBTQ+ community in particular in this case.[13]

The second question relates to the behaviour of our colleagues. Peer pressure is a powerful way of bringing about change in organisations. If we realise that our behaviour is out of line with those around us and that it is disapproved of, we are more likely to make a change: this could be to tone down some of the behaviours that people find unacceptable or possibly even to decide to leave the organisation. It should be noted that where homophobic jokes and comments are common in an organisation more people are likely to join in because they think that this is a way of being accepted.

The easier the change in behaviour is, the more likely it is that people will engage with it. In this case providing training to teams on how to draw attention to inappropriate behaviour, how to challenge effectively and demonstrating how straightforward some of these techniques are will all encourage people to adopt them.

Where people can see something in it for them, where they know that their colleagues will disapprove if they engage in such behaviour, and where the change itself is easy then people are more likely to adopt those actions.

Achieving Change in International Locations

As was discussed in Chapter 8 there is a wide variation both in terms of inclusive attitudes towards LGBTQ+ and legislation. Since the early modern period there has been this constant tension between areas of the world which are accepting of people who are in a sexual minority and those who find their behaviour, and them, objectionable. Where an organisation has its headquarters in the part of the world which is accepting, but they have regional offices in areas of the world which are best described as homophobic, it can be difficult to determine what the overall approach should be.

The same tensions can exist within a country when dealing with business locations where they are less inclusive when it comes to LGBTQ+. There needs to be an appeal to create an inclusive workplace and to recognise that most people want this. In order to be persuasive in encouraging people to look at their behaviour it's important that the request comes from people who share an identity with them and who they respect.[14] Using leaders from headquarters will have some impact but not as much as having local leaders endorsing and promoting the need to have inclusion. It is also important that the policymakers in head office don't just take a rather condescending view of those in different countries. Educating oneself about the history of LGBTQ+ is important because it means that those organisations in the West can understand the context for homophobic attitudes and legislation in different parts of the world. In many of those regions it was people from the West who criticised the tolerant attitudes that these countries had adopted and forced them to change their attitudes. It is hugely ironic therefore that those same countries should now find themselves being berated for adopting the attitudes of those from the West.

Organisations implicitly are using three models which they move between: the "When in Rome" model, the "Embassy" model, and the "Advocate" model.[15] The first model accepts the local norms and attitudes. Challenging these would be difficult and it is probably felt would achieve little. In this model, when talking about global policy, for example, little or no reference would be made to LGBTQ+. In the countries where this model is applied the government is likely to punish organisations that seek to create cultures that are more LGBTQ+ inclusive. We can only speculate the impact this will have on staff in those countries from the LGBTQ+ but there is, understandably, little research on this.

The "Embassy" model seeks to create an environment within the organisation which is more LGBTQ+ inclusive. Organisations will seek to differentiate themselves from the local culture which may be more hostile and make it clear to the staff that they want them to feel more comfortable when at work. As the authors of the research noted this is more likely an option in those regions which are not overtly hostile and so there is a degree of flexibility in what organisations are able to do.

The "Advocate" model is where organisations seek actively to change local laws for example and use their authority and status in the location to influence governments. Another way of providing advocacy, by the suppliers an organisation chooses to work with, was demonstrated by Deutsche Bank in 2019 when they announced that they had

> removed the Dorchester Collection hotel group from its list of suppliers in support of lesbian, gay, bisexual, transgender, inter and queer (LGBTIQ) rights. The decision follows the introduction of new laws against homosexuality by Brunei, whose state-owned investment agency owns the Dorchester Collection.[16]

Sensitivity needs to be encouraged in the advocacy approach because it would be hugely dangerous for someone in those countries to be outed inadvertently. However, it is important for LGBTQ+ employees in these locations to understand that there are others in the organisation who recognise the dilemmas that they are faced with and are supportive of them.

Pronouns

There are a number of decisions being made by businesses that feel like the right thing to do at the moment given what we know. These decisions are mostly being made with good intention, but we need to consider broader implications.

Over recent years, many organisations have taken steps to become more inclusive of transgender and non-binary colleagues by introducing conversations about gender pronouns.

The Oxford English Dictionary defines pronouns as "a word that is used instead of a noun or noun phrase, for example 'it, them, me, he, hers'." A gendered pronoun is one which associates a gender with the individual being discussed, so "he" or "she" for example. They typically take the place of a person's name in a sentence when we are referring to them.

We use gendered pronouns to refer to other people perpetually, and often without any thought or conscious intervention. Our brain subconsciously selects the gender pronoun based on a rapid and automatic assessment of the individual based on the information we have about them. For example, we may assume the gender of the person we are in contact with based on factors such as their name and bodily characteristics like physical size, voice, hair patterns, facial features and body shape, among many others.

However, it's important to remember that gender identity is not visible—it's an internal sense of one's own identity. While most people align across their birth-assigned sex, their gender identity, their gender expression and how everyone else interprets their gender, this isn't true for everyone.

Transgender and non-binary people may, as part of their transition or gender affirmation, ask to be referred to using pronouns that differ from those that are commonly assumed, or those that had previously being used. For example, somebody who was assigned the sex male at birth but who identifies as female may change their name and ask to be referred to using the pronouns "she" and "her." Somebody who does not identify as male or female may ask to be referred to as "they" and "them."

Referring to a person using incorrect pronouns, that is, assuming a gender that they don't identify with, is called misgendering. Using incorrect pronouns to refer to somebody may be an automatic action that is not intended to be harmful, but it does have a negative impact on the individual, as we explored in Chapter 5. Misgendering has many negative side effects for an individual, such as psychological distress, depression, reduced self-esteem, as well as feelings of being invisible, disrespected and unsafe.

As a result of increased awareness of this harmful experience, many organisations have taken steps to reduce the gendered language in their policies and documentation, as well as in their spoken communication. Workplace policy documentation has historically been peppered with the pronoun "he" to refer to any employee, and over time this became replaced with he/she. We now see the pronoun "they" being used to refer to employees in this situation.

Organisations have also begun incorporating questions about pronouns into recruitment and onboarding processes, as well as internal databases.

The biggest change we have seen in organisations over the past few years is the move towards making pronouns visible. You may have noticed others sharing their pronouns in various ways for example on their email footers, business cards, name badges and social media. Some organisations have also adopted pronoun policies in meetings, whereby introductions should include an individual's pronouns.

We have found that some organisations have mandated the sharing of pronouns and a commonly cited reason for doing this is to ensure transgender and non-binary employees are not being singled out. Where this occurs, transgender and non-binary employees are more likely to feel comfortable talking about pronouns and they are less likely to be misgendered. Another benefit is that this raises awareness of pronouns, giving people the opportunity to learn more about how to make others feel included by using appropriate and respectful language.

However, visibly sharing pronouns can also have some unintended and negative consequences.

It is essential that we take a step back to look at some of these considerations in order to ensure that we are pursuing the most effective and inclusive methods and reducing any harmful consequences of our actions.

The first thing that we must consider is whether this public and visible sharing of pronouns is inclusive for all members of the transgender and non-binary community, including those who do not disclose their identity to others.

In their 2018 report investigating the workplace experience of LGBT employees, Stonewall found that 51% of transgender respondents had hidden their identity at work in the previous year due to concerns about discrimination[17] The report further showed that one in four trans people (26 per cent) aren't open with anyone at work about being trans. This number increases to almost two in five non-binary people (37 per cent) who aren't out at work.

Concealment rates within this community are high, and as we explored in Chapter 4, there are various reasons for this. The most commonly cited reason for concealment of transgender identity is concern regarding discrimination.

It's also important to highlight here that these figures are reflective of those who volunteered to take part in the research, and therefore it is likely that there are in fact many other individuals who conceal their identity and are not comfortable participating in research, or who are still coming to terms with their own identity, who would not be included in this reporting. This number is therefore likely to be an underrepresentation of the true figures.

For those individuals who conceal their gender identity from others at work, the sharing of gender pronouns is a lose-lose situation. On the one hand, they can share their correct pronouns but in the process will have outed themselves, potentially exposing·themselves to discrimination, bullying, harassment and exclusion by others. Or they can share incorrect pronouns and continue to conceal their identity.

The impact here is that the visibility of these pronouns serves as a persistent reminder to the individual of the disparity between their identity and how others see them. Further, publicly sharing these incorrect pronouns means that others will continue to use them, perhaps even more so because they feel this is a confirmation of identity, and therefore they may be subject to misgendering on a more frequent basis.

This constant reminder is likely to enhance feelings of gender dysphoria that is experienced by some members of the transgender and non-binary community. This leads to anxiety, social withdrawal and cognitive preoccupation. All of these side effects not only impact an individual's well-being but also impede an individual's performance.

The second challenge that is posed by visible pronouns relates to the activation of gender stereotypes. When aspects of our identity are made salient, stereotypes are more readily activated. It would therefore follow that when we see pronouns such as he/him or she/her, the stereotypes that we associate with gender will come to mind.

Stereotypes of men and masculinity include being strong, skillful, competent and intelligent, whereas the stereotypes of women and femininity include being friendly, warm, kind and nice.

These stereotypes are both descriptive and prescriptive, which means that they generate expectations about how men and women should be, as well as how they are. The act of reminding somebody, or indeed simply reminding yourself, of your gender can increase the likelihood that stereotypes will be drawn upon in subsequent interactions.

However, discrimination from other people is not the only way that stereotypes can impact our careers; they can also influence the way that we ourselves interpret our own abilities. Stereotype threat occurs when an individual feels they are at risk of confirming a negative stereotype about themselves. Worrying about conforming to stereotypes that others have about you based on things like gender has a cognitive effect, which often then impacts the individual's ability to perform to the standard they are capable of. It can also lead to career limiting decisions for example about what roles to apply for and career development to engage in.

The research in this domain consistently highlights the factors that make gender more salient heighten that association with gender stereotypes, and as a consequence, exaggerate the perceived differences between men and women and the resulting discrimination that we see in the workplace.

We recommend that employees have the option to share their pronouns in some of the ways we have discussed, if they feel comfortable doing so. Don't mandate this move, as it's more likely to make the most vulnerable within the community feel further excluded.

It is these steps that will ensure that we don't exclude those who conceal their identity, and we don't risk increasing the likelihood of activating stereotypes. For example, this should include ensuring employees are aware of the importance of using correct pronouns and the impact of misgendering colleagues. Outside of this, the role of gender stereotypes and the impact of bias should be a consideration for organisations, again through training and awareness raising. Finally, reinforcing to everyone that discrimination of and harassment towards transgender and non-binary colleagues are unacceptable will help employees to feel more comfortable sharing their identity without the threat of negative behaviours from colleagues.

Toilets

"Something which has never occurred since time immemorial; a young woman did not fart in her husband's lap." Whatever you think about this joke, it is, according to the University of Wolverhampton, the world's oldest, having been found on Sumerian tablets from the old Babylonian period, from around 2300 BCE.[18] It also shows the enduring appeal of toilet humour, which relies upon talking publicly about aspects of bodily functioning which are normally considered impolite and even taboo.

Toilet provision in organisations is something that can be trivialised and yet there are serious and substantive issues to be considered.

Public toilets have a gendered as well as a racial history.[19] In the United Kingdom in the Victorian era public toilets were available for men but not for women—as their place was seen to be in the home. The first public provision of toilets took place at the Great Exhibition of 1851 at the Crystal Palace. Their success led to the City of London installing more public conveniences beginning in 1855. It wasn't until another nearly 40 years had passed before toilets were made available for women.[20] In other words, toilets were a feminist issue. The same thinking applied in the USA where public toilets for women were not provided until the 20th century. City centres and towns became places where men felt more comfortable. Furthermore, in the south of the United States toilets were racially segregated and far less clean for minorities than they were for the white population.

It's easy to see why discussion about toilets could be uncomfortable and not taken seriously. It is also easy to shrug one's shoulders and roll one's eyes when people discuss this topic, as if it is of little significance compared to the major issues that an organisation faces.

Nevertheless, it needs to be recognised that this has always been an area which has been contested particularly when it comes to the provision of amenities for women. It should be no surprise then that it is a subject that arouses much emotion within organisations. The United Nations has identified sanitation as being important for the achievement of gender equality, one of its Sustainable Development Goals. A framework was developed so that consistent standards for sanitation could be applied globally, and they included the need for "availability; affordability; quality and safety; acceptability, privacy, and dignity."[21]

Historically, toilets within public areas and offices have been divided into three distinct spaces: spaces for men, spaces for women, and accessible spaces for disabled individuals of any sex or gender. Over the last decade, there has been a significant amount of pressure on businesses to allow people to use the space that feels most appropriate for them based on their own gender identity. So, for example, a transgender man would have access to the men's toilets, and a transgender woman would have access to the women's toilets. Many organisations have since adopted such a policy, some taking things even further by ensuring all toilet spaces are "gender neutral." However, there are some broader implications that we have to consider, such as the safety of individuals in using such spaces, and the risk of increases in sexual assault.

Women have concerns about safety not just in using the facilities but on approaching them and on exiting them.[22] It is not sufficient to say that these concerns are of no relevance when it comes to the provision of gender-neutral toilets. It also needs to be appreciated that transgender and non-binary individuals are not necessarily comfortable or safe in using such facilities either. Nearly half of transgender people (48%) are fearful of violence they may experience because of using public toilets.[23]

Providing gender-neutral facilities therefore is not as straightforward as it might appear. There are a number of issues that need to be resolved usefully summarised by Gail Ramster and her colleagues:

> the level of privacy afforded to users of cubicles; the appropriateness of gender-neutral facilities with urinals and how users and nonusers might feel excluded from this space; the potential for confusion and anxiety amongst some users of a previously familiar facility; and the current threatening architecture of some public toilet facilities. (p111)[24]

It's because of this that building regulations have begun to shift again away from this rise in gender-neutral toilets.[25] In an ideal world, all toilet spaces should be gender-neutral and fully enclosed with locks (i.e., single cubicles). But just how practical is that for an organisation with a fixed office size? We want to have privacy when using toilet facilities and have come to accept sharing with those of the same gender. It is a widely accepted cultural practice, and asking people to change this is something that will meet with resistance because of the discomfort that people experience. We need to be careful, as some researchers have pointed out, that we don't create "new routes to exclusion" (p112).[26] Care and caution need to be exercised when considering changes to toilet facilities. It is essential that organisations listen to the needs of all of the staff and seek the advice of a representative range of professional advisors to reach a solution which can accommodate the needs of everyone.

Communication/Visibility

Communication should happen on two levels. Across the organisation, whether you are in a management, HR or a D&I role, be proactive about your LGBTQ+ support and not reactive. Make sure you are aware of important dates in the LGBTQ+ calendar such as Pride Month, LGBT history month and days that are internationally recognised to stand against homophobia, biphobia and transphobia. Communicate your support widely and regularly—don't make it a one-off or limit it to these events.

Be consistent in your approach and prove to your employees that this is not just another example of a pride flag on your Twitter logo with no tangible support behind it. Showing your support throughout the year and in different ways gives your actions greater meaning and demonstrates a deeper level of commitment which will have an impact on your clients and customers as well as your colleagues.

The Confronting Prejudiced Responses model describes a helpful set of steps to create a truly inclusive organisation.[27] It provides a very useful way of looking at how people can intervene—not just in terms of the actions themselves but also in the work that needs to happen beforehand in order to create an environment where people feel safe enough to act. These include:

1. Detecting discrimination-recognising the different forms of discrimination can take. Most people can identify the blatant forms, but the subtle behaviours can escape many people unless you're on the receiving end of them.
2. Recognising that the incident is an "emergency" (p305). The language is rather dramatic but it does emphasise the fact that many people can underestimate the significance that discriminatory behaviour can have on individuals. They can explain it away by saying "it wasn't that serious," "they didn't mean it," and "you're overreacting"—all of which will mean that the incident is not being taken seriously. Education and awareness raising will help, particularly in helping to increase understanding about the impact that such behaviour has on people's long-term well-being is also necessary.
3. Taking responsibility to take action. In situations where the behaviour is ambiguous there will be a tendency to look around at how other people are reacting. That will then determine how we respond. If, however, everyone else is in the same position and looking at us to see how we react, the outcome will be that nobody does anything. So, in order to be an effective ally, people need to understand how important it is we take responsibility for acting in that situation
4. Knowing the most effective action to take. It can often be assumed that there is only one way of challenging people, that is, to do it very directly and assertively. That certainly is one approach, but others are available too and people need to know the circumstances. This is where training is important so that people understand that not only can we take action after something has happened, but we can also create environments where discriminatory behaviour is less likely to occur in the first place. It has become more routine for people to say that for example "micro aggressions need to be called out." This suggests that there needs to be a public demonstration of the challenge so that people can see that something is unacceptable. However, sometimes, perhaps often, the more sensitive approach needs to be adopted where people are given the feedback, but it is done privately. Publicly calling out people is more likely to result in people feeling ashamed. This in turn will mean that they are less likely to engage in anything related to diversity and inclusion in the future. Instead, depending on the behaviour obviously, it is far better to provide feedback to people in a way which enables them to reflect and develop.

In Conclusion

Understanding the experiences of LGBTQ+ individuals in the workplace is essential for creating a truly inclusive environment. Reflecting on ancient times, where homophobia was virtually non-existent and people were accepted for who they were, provides a powerful reminder of what is possible. By recognising the fluidity and diversity of identities that existed in the past, we can appreciate that rigid labels are neither natural nor necessary.

Today's workplace has the potential to be a space where labels do not matter and where everyone is valued for their unique contributions. Acceptance of different labels is crucial, but we must go beyond aspects such as rainbow lanyards and Pride Month celebrations, as important as they are. True inclusivity requires us to go beyond actions such as these; it demands a deep, sustained commitment to valuing every individual.

Creating such a world involves challenging our own biases, advocating for inclusive policies and cultivating environments where authenticity is celebrated. Each of us has a part to play in fostering truly inclusive LGBTQ+ workplaces, which requires us to challenge our own attitudes as well as those of others. Embracing multiple identities allows us to move beyond binary thinking and reduces the emphasis on LGBTQ+ being seen as a defining characteristic. The shifting perspective can help us focus on commonalities rather than differences, fostering a culture of acceptance and mutual respect.

In this vision of the future, the significance of an LGBTQ+ identity will diminish, replaced by a recognition of our shared humanity. By embracing diversity in all its forms, we not only honour the legacy of acceptance from the past but also pave the way for a more equitable and compassionate world.

Notes

1 Hoel, H., Lewis, D., & Einarsdóttir, A. (2014). *The Ups and Downs of LGBs' Workplace Experiences*. Manchester Business School.
2 Bachmann, C.L., & Gooch, B. (2018). *LGBT in Britain: Work Report*. Stonewall.
3 TUC. www.tuc.org.uk/sites/default/files/LGBT_Sexual_Harassment_Report_0.pdf
4 https://coqual.org/reports/the-power-of-out-2-0/
5 https://coqual.org/reports/the-power-of-out-2-0/
6 Kidd, D.C., & Castano, E. (2013). Reading Literary Fiction Improves Theory of Mind. *Science*, 342, pp. 377–380.
7 www.shrm.org/topics-tools/news/ergs-pass-deloitte-phasing-inclusion-councils
8 www.bloomberg.com/news/articles/2017-07-19/deloitte-thinks-diversity-groups-are-pass?embedded-checkout=true
9 Einarsdóttir, A., Mumford, K., Birks, Y., Lockyer, B., & Sayli, M. (2020). *Understanding LGBT+ Employee Networks and How to Support Them*. University of York.
10 Calabrese, S.K., Earnshaw, V.A., Magnus, M., Hansen, N.B., Krakower, D.S., Underhill, K., Mayer, K.H., Kershaw, T.S., Betancourt, J.R., & Dovidio, J.F. (2018). Sexual Stereotypes Ascribed to Black Men Who Have Sex with Men: An Intersectional Analysis. *Archives of Sexual Behavior*, 47(1), p. 143. Accessed 16th June 2022.

11 Goldstein, S.B., & Davis, D.S. (2010). Heterosexual Allies: A Descriptive Profile. *Equity & Excellence in Education*, 43(4), pp. 478–494.

12 Ajzen, I., 1991. The Theory of planned behavior. *Organizational Behavior and Human Decision Processes*.

13 Donovan, P. (2020). *Profit and Prejudice: The Luddites of the Fourth Industrial Revolution* (1st ed.). Routledge. https://doi.org/10.4324/9781003098898

14 Vasilev, G. (2016). LGBT Recognition in EU Accession States: How Identification with Europe Enhances the Transformative Power of Discourse. *Review of International Studies*, 42(4), pp. 748–772.

15 Glasgow, D., & Twaronite, K. (2019). How Multinationals Can Help Advance LGBT Inclusion Around the World. *Harvard Business Review*. https://hbr. org/2019/08/ how-multinationals-can-help-advance-lgbt-inclusion-aroundthe-world

16 www.db.com/news/detail/20190404-deutsche-bank-has-removed-all-hotels-owned-by-the-sultanate-of-brunei-from-its-supplier-list?language_id=1#:~:text=Diversity%2C%20Equity%20%26%20Inclusion-,Deutsche%20Bank%20has%20removed%20all%20hotels%20owned%20by%20the%20Sultanate,and%20queer%20(LGBTIQ)%20rights

17 Bachmann, C.L., & Gooch, B. (2018). *LGBT in Britain: Trans Report*. Stonewall.

18 www.wlv.ac.uk/news-and-events/latest-news/2008/august-2008/the-worlds-ten-oldest-jokes-revealed.php

19 Lewkowitz, S., & Gilliland, J. (2024). A Feminist Critical Analysis of Public Toilets and Gender: A Systematic Review. *Urban Affairs Review*, p. 10780874241233529.

20 Ramster, G., Greed, C., & Bichard, J.A. (2018). How Inclusion Can Exclude: The Case of Public Toilet Provision for Women. *Built Environment*, 44(1), pp. 52–76.

21 Lewkowitz, S., & Gilliland, J. (2024). A Feminist Critical Analysis of Public Toilets and Gender: A Systematic Review. *Urban Affairs Review*, p. 10780874241233529.

22 Hartigan, S.M., Bonnet, K., Chisholm, L., Kowalik, C., Dmochowski, R.R., Schlundt, D., & Reynolds, W.S. (2020). Why Do Women Not Use the Bathroom? Women's Attitudes and Beliefs on Using Public Restrooms. *International Journal of Environmental Research and Public Health*, 17(6), p. 2053.

23 Ramster, G., Greed, C., & Bichard, J.A. (2018). How Inclusion Can Exclude: The Case of Public Toilet Provision for Women. *Built Environment*, 44(1), pp. 52–76.

24 Ramster, G., Greed, C., & Bichard, J.A. (2018). How Inclusion Can Exclude: The Case of Public Toilet Provision for Women. *Built Environment*, 44(1), pp. 52–76.

25 www.cibsejournal.com/technical/building-regulations-update-single-sex-toilets/

26 Ramster, G., Greed, C., & Bichard, J.A. (2018). How Inclusion Can Exclude: The Case of Public Toilet Provision for Women. *Built Environment*, 44(1), pp. 52–76.

27 Ashburn-Nardo, L., Morris, K.A., & Goodwin, S.A. (2008). The Confronting Prejudiced Responses (CPR) Model: Applying CPR in Organizations. *Academy of Management Learning & Education*, 7(3), pp. 332–342.

Index

For Product Safety Concerns and Information please contact our EU
representative GPSR@taylorandfrancis.com
Taylor & Francis Verlag GmbH, Kaufingerstraße 24, 80331 München, Germany

www.ingramcontent.com/pod-product-compliance
Ingram Content Group UK Ltd.
Pitfield, Milton Keynes, MK11 3LW, UK
UKHW041932250625
460073UK00008B/119